APPLES

& HOW TO GROW THEM

A COMPREHENSIVE GUIDE TO 400 APPLE VARIETIES WITH PRACTICAL TIPS FOR GROWING, HARVESTING AND STORING

Andrew Mikolajski

Photography by Peter Anderson

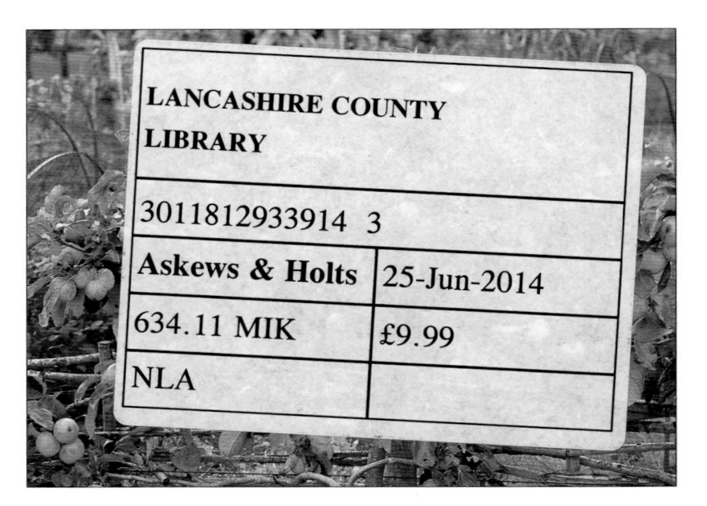

This edition is published by Southwater
an imprint of Anness Publishing Ltd
108 Great Russell Street
London WC1B 3NA
info@anness.com

www.southwaterbooks.com
www.annesspublishing.com

If you like the images in this book and would like to investigate
using them for publishing, promotions or advertising, please visit
our website www.practicalpictures.com for more information.

A CIP catalogue record for this book is available from the British Library.

Publisher: Joanna Lorenz
Project Editor: Anne Hildyard
Photographer: Peter Anderson
Illustrators: Maj Jackson-Carter and Liz Pepperell
Designer: Nigel Partridge
Production Controller: Wendy Lawson

Previously published as part of a larger volume, *The Illustrated World Encyclopedia of Apples*

PUBLISHER'S NOTE

Although the advice and information in this book are believed to be accurate and true at the time of going to press, neither the authors nor the publisher can accept any legal responsibility or liability for any errors or omissions that may have been made nor for any inaccuracies nor for any loss, harm or injury that comes about from following instructions or advice in this book.

ACKNOWLEDGEMENTS

The publisher would like to thank the following for their contribution to this project:
Jim Arbury at RHS Garden Wisley, Surrey, UK; and Brogdale Collections, Kent, England.
Picture credits: Alamy 13tm; Corbis 12tr; GAP Photos 17t, 18tl, 21b, 24tm.

CONTENTS

INTRODUCTION

The apple has long been one of the most popular – if not the most popular – of all fruits and much time and energy have been devoted to its development. Prized for its taste and keeping qualities, the humble apple is now the most widely cultivated tree fruit, both in commerce and in gardens.

No one knows exactly when apples were first discovered and eaten, but they are believed to have originated in Turkey. The fruit would have been taken via ancient trade routes to virtually every part of the huge land mass that comprises Eurasia. In more modern times, trees have been introduced into the Americas, southern Africa and Australasia. Nowadays they are grown throughout temperate parts of the globe and have become of great significance to the economies of many countries as an exportable commodity. Such is the adaptability of the apple, that it is only in tropical, desert and extremely cold regions that trees cannot be relied on to produce good crops.

Due to their ease of cultivation and their tolerance of changeable climates, apples have never been considered a luxury food. Historically, they were an essential element in the diet of many peoples from all strata of society.

Above: Apple blossom is a captivating sight in the spring garden. This is the crab apple Malus x schiedeckeri *'Hillieri'.*

A fruit for health and fitness

The apple has often been seen as a symbol of youth, fertility and longevity. It is frequently offered as a love token between men and women who are of marriageable age.

Left: Hand picking ripe apples can be enjoyed by all the family.

'An apple a day keeps the doctor away' is an old proverb and recent research suggests that there may well be more than a grain of truth in this. The fruit contains natural substances that studies indicate may be able to play a significant part in strengthening bones and lowering cholesterol as well as providing some protection against certain cancers, Alzheimer's disease, asthma and other respiratory diseases. This is aside from their natural vitamin content and the dietary fibre they provide, both essential in a balanced diet. Apples are the ultimate convenience food, small enough to slip into a lunch box (or even your pocket) and providing an ideal snack when you are on the move and hunger strikes.

A versatile fruit

Down the centuries, apples have been bred for improved size and flavour and, very importantly, for storing for use when fresh fruits and vegetables are in short supply. Not only are the fruits delicious eaten straight from the tree but, unlike many other fruits, they can be transported over long distances or kept for many months without any adverse effects on their flavour.

Besides the types that are grown for eating raw, the vast range of varieties – running into thousands – includes apples for cooking, for juicing and for cider making. In the kitchen, fruits can be used both for sweet and savoury dishes; the tart, acid flavour of some is a perfect complement to many rich or fatty meats, especially pork and game, while others have a flavour as sweet as strawberries or pineapples.

Nowadays, with international trade and improvements in storage, there is not a single day of the year when a tasty apple cannot be enjoyed. Walk into any supermarket and you will have a choice of several varieties. A farm shop or farmers' market, where locally produced foods are sold, may offer an even wider selection.

Easy to grow

Hardy and woody as plants, apples are supremely well adapted to growing in domestic gardens. Unlike many fruit trees, they do not necessarily need a lot of room – though a large, mature apple tree can be a magnificent sight. On dwarfing rootstocks or trained on wires in a potager or on an allotment, apple plants take up very little space but can still produce abundant crops. Many have been bred for disease-resistance and need only the minimum of maintenance once established. Some varieties can be grown in tubs or large containers, so even if you have only a patio or balcony, you can still have fresh fruit within easy reach.

How to use the directory

The directory section is arranged alphabetically. Details of over 400 apples are given, ranging from old cultivars from major apple growing areas, some now rare in cultivation but found in historic orchards and gardens, alongside more modern cultivars that are found in supermarkets and garden centres. The whole fruits are shown, with, in most cases, the cut fruit alongside, and sometimes with an additional photograph of the fruit on the tree.

Each entry comprises a description of the fruit when ripe: its shape, size, and skin colour, followed by the colour and texture of the flesh and an indication of the flavour.

Shapes of older varieties can vary, some being of very uneven appearance. These varieties are generally less widely grown today but are nevertheless of considerable interest. Sizes of fruits are necessarily approximate. While modern varieties have been bred for uniformity, older varieties in particular can carry fruits of very different sizes that can ripen at the same time. Note also that the taste of a fruit is subjective and to some degree depends on where the apple was grown, the soil type, when it was picked and how long it has been stored.

The information panel indicates the type of apple – dessert, for eating raw, or culinary, for cooking – and its origin: the place where and date when it was first grown. Dates of introduction are often of necessity approximations, even for more modern cultivars which may be extensively trialled before they are available in commerce, so may not always accurately reflect the date the variety was bred. Some are known to have been grown in gardens for many years before being grown commercially. Equally, it is only in more recent times that certain eastern European varieties have become more familiar with the growth of international trade throughout continental Europe.

The parentage of the apple, where known, is given. This can often provide hints as to the flavour of the apple and its performance in gardens.

Conjectural details concerning date, place of origin, and parentage appear with question marks (?) before them.

The flowering season provides an indication of when the flowers will be open – important when choosing compatible cultivars for cross-pollination. However, flowering times may vary from year to year as this can depend on the length and severity of the preceding winter.

Where they exist, other names of the variety are also listed. Some varieties go under different names in different countries. Older apples, bred before the naming of plants was strictly regulated, have often been grown under a number of names.

The note indicates special characteristics, such as resistance to disease, rate of growth, length of time the fruits can be stored, or any other detail of interest. Cultivation is assumed to be broadly similar for all varieties, but where a particular apple is known to do well in certain conditions this is indicated.

Above: An apple tree provides a pretty, tranquil spot for rest and contemplation.

Identifying varieties

This book aims to bring a greater understanding and appreciation of the huge variety of apples, all aspects of cultivating and propagating them as well as any potential problems that may affect apple growth.

While the directory is to some extent intended as a diagnostic tool, it may not be possible to identify an unknown apple through referring to it alone, bearing in mind the vast number of cultivars in existence. Reflecting the popularity of the fruit and the growing interest in historic gardens and home-grown produce, many botanic gardens host special apple days in autumn, when it is possible to have fruits identified by experts.

Below: Harvesting and storing your apples is immensely rewarding and satisfying

GROWING APPLES

Apples have been cultivated for hundreds of years, but with modern cultivation methods, you don't need large amounts of space in which to grow your own fruit trees.

It is possible to train trees against walls and fences or use them as garden dividers, and there are even dwarf forms of popular fruits that are small enough to grow in containers. Established fruit trees are able to compete with other plants for moisture and nutrients and so can be accommodated all around the garden. Make sure they are planted in a sunny spot and that there is sufficient access to carry out essential maintenance. If you want to grow a lot of fruit, you should consider allocating a separate area of the garden where the fruit trees are easier to manage and protect.

The following pages will guide you through the satisfying process of growing your own apples, from choosing a tree and planting to pruning, nurturing and after-care.

***Above left**: These pole apples have been grown as vertical cordons, which are perfect for the kitchen garden where space is limited.*
***Above right**: The shiny, bright red fruit of* Malus pumila *'Dartmouth'.*
***Left**: When apple trees are in full blossom they make a stunning display.*

WHAT IS AN APPLE?

Apples belong to the genus Malus, *which comprises some 35 species.* Malus *is a member of the Maloideae subfamily of the family Rosaceae – the rosaceous plants – and apples are thus related to such common garden plants as roses (*Rosa*), cotoneasters and rowans (*Sorbus*) as well as other plants grown for their fruits.*

Nowadays, cultivated apples are all grouped as *Malus domestica*, not a true species but also not a chance hybrid as was previously assumed.

Apples have always been the most widely grown fruit in the temperate zone, with pears a somewhat distant second. Apples are also grown in Mediterranean regions and in cooler areas on higher ground in the tropics.

Geographic origin

While there are a number of trees with apple-like fruits found growing in the wild, the ancestors of the apples we enjoy today are believed to derive from *Malus sieversii*, a species that is found in the mountains of Central Asia in southern Kazakhstan, Kyrgyzstan, Tajikistan and Xingjian, China. This has the largest fruit of any wild apple, up to 7cm (2½in) in diameter. DNA taken from a tree growing in the Ili Valley, at the border of northwest China and Kazakhstan, shows some genetic sequences that are common to *Malus domestica*.

M. sylvestris, a European species found in a range from as far south as Spain, Italy and Greece to Scandinavia and parts of northern Russia, has also contributed to the genome of the cultivated apple, but to a lesser extent than has previously been assumed. *Malus baccata* (found in central Japan and central China) may also have played a part in the development of some varieties.

The plant

An apple plant is a deciduous tree or large shrub. The species from which modern varieties have been bred is a tree of average (though variable) size, at most 15m (50ft) in height. Because of extensive interbreeding, it is difficult to state precisely how big any particular variety can grow. Most are kept artificially small through the use of rootstocks and pruning.

Apple plants begin to bear fruit when they are around four or five years old, and trees have been known to remain productive for more than 200 years.

Flowers

Along with other members of the rose family, apples have single, cup-shaped, five-petalled flowers that open flat to up to 5cm (2in) across. They are usually white, some flowers having pink markings. Some species that are grown as ornamentals, often called crab apples, have deep pink flowers.

Leaves

Apple trees have simple leaves that are arranged alternately along the length of the stem. The leaves are bright to dark green in colour, with toothed margins. The undersides of most apple leaves are greyish-silver in colour and slightly downy. Leaves turn various shades of yellow, orange and red before falling from the tree during the autumn.

Above: Skin colours range from deep red through to golden yellow or dark green. The seed chambers are centred around the core.

Fruits and seeds

The fruit of the apple is classified as a pome. Fruits are round or slightly elongated and narrow slightly towards the blossom end, opposite the stalk.

Ripe fruits have skins that are red, yellow or light green. In red-skinned varieties, the red colour develops only as the fruit ripens, overlaying yellow or yellowish-green. In some varieties, the red almost masks the base colour entirely, while on others it appears as attractive striping or light flecking. Even on ripe fruits that appear evenly red, there will be a trace of green at the core, where the stalk meets the fruit. Apple skins can be waxy, with a slight shine, or show dull, rough patches known as russeting.

Above: The flowers of the apple tree are usually white, sometimes flushed with pink.

Above: The topside of an apple leaf is a deep green colour.

Above: The underside of an apple leaf is a green- or silvery-grey.

Growth cycle

In mid-spring, flower buds emerge simultaneously with the leaves. The unopened flower is encased within five green sepals. These split and curve back to reveal five petals.

In their search for nectar and pollen, bees transfer pollen from the anthers (male) of one flower to the stigmas (female) of another. Pollination occurs when pollen forms a pollen tube that grows downwards to the ovary. In the ovule, a male cell fuses with a female (fertilization).

The petals fall and the style wilts. The ovary grows to become a fruit, covering the fertile seeds within it. Seeds comprise an embryo plant inside a hard outer coat. The seeds mature as the ovary expands to form a ripe fruit.

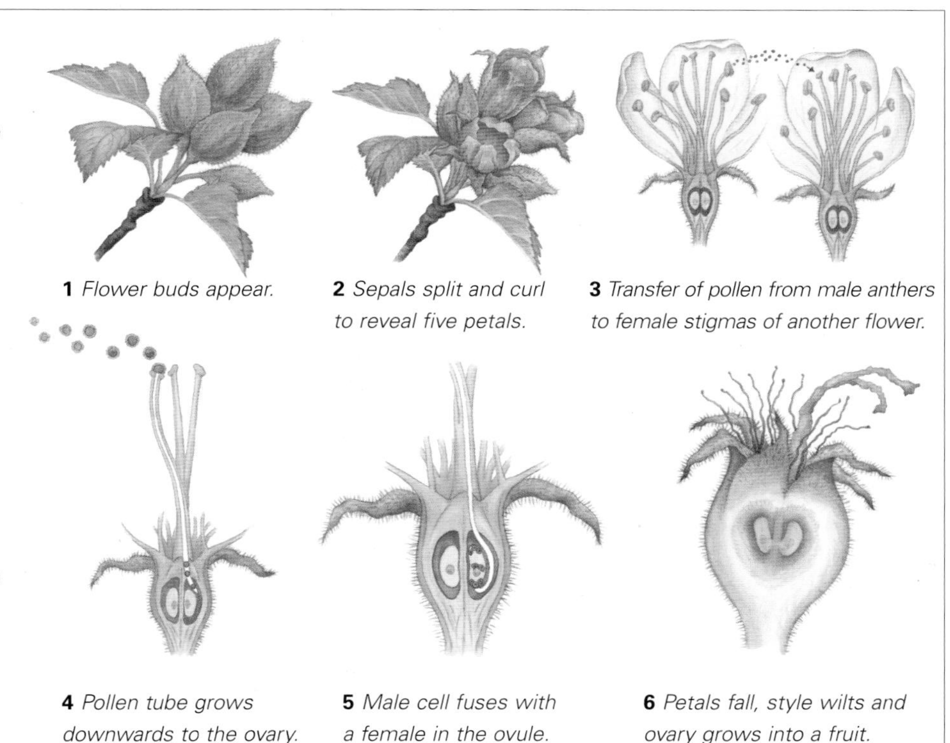

1 *Flower buds appear.*

2 *Sepals split and curl to reveal five petals.*

3 *Transfer of pollen from male anthers to female stigmas of another flower.*

4 *Pollen tube grows downwards to the ovary.*

5 *Male cell fuses with a female in the ovule.*

6 *Petals fall, style wilts and ovary grows into a fruit.*

Like other members of Rosaceae, the seeds develop inside a leathery core. Cutting an apple in half cross-ways reveals a star with five chambers, each containing two seeds. Seeds are dark brown when ripe.

The texture of the flesh is firm, hard when unripe, and mealy in some varieties but usually remaining crisp. It varies in colour from white to pale creamy-yellow. Freshly cut, the exposed flesh rapidly oxidizes, turning yellowish-brown on contact with the air. All parts of the ripe fruit are edible. Other edible pome fruits include pears (*Pyrus communis*) and quince (*Cydonia oblonga*).

Crab apples

A number of *Malus* species are grown mainly as ornamentals. Crab apples make attractive specimens in small gardens, having a generally dainty appearance. They also attract beneficial insects and birds into the garden and can act as pollinators for cultivated apples.

White or pink spring flowers are followed by decorative autumn fruits and, often, good leaf colour. Though the fruit, which is much smaller than cultivated apples, is edible, it usually has to be cooked to make it palatable. Crab apples are popularly used in jellies and jams.

Cross-pollination

When a bee transfers pollen from the stigmas of one flower to those of an open flower on another tree, this is cross-pollination.

Most apples are self-sterile, that is to say that the transfer of pollen from one open flower to another on the same tree will not result in fertilization, and hence fruit. So, for apples to crop successfully, it is necessary to grow two or more trees near to each other that flower at the same time.

The function of fruit

Edible fruit is designed to be attractive to passing animals that eat the fruit, and, in the process, the seed. The indigestible seed is then carried in the animal's gut some distance from the parent plant before being eliminated in its droppings. In favourable conditions, the seed will germinate and grow into a new tree. This ensures that the genetic potential of the parent plant will be spread over an increasingly large area with each succeeding generation.

Depending on their flavour, apples are classed as either dessert (for eating raw) or culinary (for cooking). Culinary apples have a tart flavour that is unpalatable until the flesh is cooked and, usually, sweetened before use. Depending on the variety, culinary apples either retain their shape during cooking, which is good for open tarts, or collapse into a fluffy mass, which is ideal for sauces and purées. Many varieties are multi-purpose and can be eaten raw or used in cooking.

Above: *The Egremont Russet is almost covered in yellowish-brown russeting.*

APPLE BREEDING AND COMMERCE

The cultivation of apples dates back to the time of the Egyptian Pharaohs, when apples were grown along the Nile Delta. By the 7th century BCE, they were known in Greece, and by the first century CE, the ancient Romans had up to 40 varieties, some of which they introduced to the western Empire, including England.

By the first century CE, apples were being grown in every region throughout the Rhine Valley, Germany. The English horticulturist John Parkinson (1567–1650) noted 60 varieties growing in England in 1640. This number had increased to 92 by 1669.

Apple growing in the Americas began in the 16th century, when Spanish invaders took the fruit to Mexico and South America. But when English settlers arrived in North America in the 17th century, they found only native crab apples.

British settler William Blaxton (1595–1675), who arrived in New England, in 1623, is credited with planting the first apple variety on American soil. He developed a nursery in Boston, and bred the first American apple – Blaxton's Yellow Sweeting.

In 1630, newly arrived Puritans took over the orchard. Apple growing spread westward into the Ohio Valley and Great Lakes areas. John Chapman, nicknamed Johnny Appleseed, from Massachusetts, became famous for planting trees throughout Ohio, Indiana and Illinois.

Below: Despite technological advances, picking apples by hand remains the best method.

Johnny Appleseed

Born John Chapman (1774–1845), Johnny Appleseed was an American pioneer nurseryman who introduced apple trees to large parts of Ohio, Indiana and Illinois.

Often depicted in rags, he became famous for his nomadic subsistence lifestyle, often going barefoot and preaching the gospel. In some ways, he was an early environmentalist, as he cared deeply about animals, even insects.

Right: A barefoot Johnny Appleseed cultivates the land before sowing seeds.

Apples were taken to the West Indies in the 18th century. By 1866, 643 different varieties were recorded in Andrew Jackson Downing's *Fruits and Fruit Trees of America.*

Apples have been grown in South Africa since 1652, when the Dutch colonial administrator Jan van Riebeeck, established the first plantings.

Orchards were established in Australia during the 1780s after apples were introduced by British Admiral Arthur Philip (1738–1814). Australia is home to the popular Granny Smith.

Diploids and triploids

Cultivated varieties of apples fall into two groups according to their chromosome count. The majority of apple varieties, along with all other living things that reproduce sexually, are diploids, with 34 chromosomes (17 from each parent). Triploids have 51 chromosomes, 34 supplied by the female and 17 by the male.

Most commercially grown apples are diploids, but there are also a large number of triploid varieties, including the well-known Jonagold and Bramley. The earliest known triploid apple is Gravenstein, grown as early as the 17th century in Denmark.

Triploids are popular because of the the large fruit that they produce and their tendency to contain more vitamin C than diploid varieties. However, triploids set fruit on a relatively small proportion of the flowers and, since they produce very little viable pollen, cannot be relied upon to pollinate other apples. So, if choosing a triploid to grow at home, you will also need to cultivate two diploid plants nearby in order to pollinate both the triploid and each other.

Polyploidy (the possession of more than the usual number of chromosomes) plays an important role in the evolution and breeding of apples.

Nutritional content

Low in calories, apples contain 80–85 per cent water. Raw, unpeeled apples are an excellent source of vitamin C and dietary fibre (both soluble and insoluble). Soluble fibre guards against a build-up of cholesterol in the blood, while insoluble fibre provides bulk in the intestines, holding water that cleanses and moves food quickly through the digestive tract. The fruit's nutritional content is not greatly affected by storage, though there may be some negligible deterioration.

Apples in commerce

While a lot of hybridizing and mutation has occurred naturally, agriculturalists, nurserymen and gardeners have grown numerous cultivars (varieties) through selective breeding programmes. Since these have never been systematically recorded, the parentage of any particular variety remains largely conjectural.

Nowadays, apples are an important commercial crop. China exports the largest number, closely followed by the USA. Turkey, France, Italy and Iran are also among leading exporters.

Repeated crossing and back-crossing has resulted in over 8,000 varieties. Of these, only around 100 are grown commercially, with ten varieties making up 90 per cent of production. A few varieties, such as Delicious and Gala, are grown throughout the world, while others are grown almost exclusively in a single location.

When choosing cultivars, growers look for reliable bearing, good flavour, attractive appearance, pest resistance, volume of crop and hardiness.

If the fruit stores and processes well, and withstands transportation, the value of the variety increases. Waxy-skinned varieties are preferred. Apples that commonly show signs of russeting are seldom grown commercially.

Harvesting

Fruits are harvested from early morning and throughout the day. However, varieties that are prone to bruising, such as Golden Delicious and Pink Lady, are picked from late morning, after the dew has evaporated.

Apples are picked by hand then collected in crates, often lined with plastic to prevent rubbing and abrasions. These crates are transported to a cold store within 24 hours.

Chemical treatments

Apples grown commercially are often treated with a chlorine drench as a phytosanitary measure to reduce the possibility of decay during storage. Golden Delicious, Braeburn and other varieties that are susceptible to diseases such as bitter pit or lenticel spot may be subjected to a calcium drench.

Above: Harvested apples are emptied into large crates to be transported by tractor.

Granny Smith, Pink Lady, Golden Delicious and other apples prone to superficial scald are sometimes treated with diphenylamine (DPA). In some instances, yeast and a fungicide are added to the drench as a means of reducing possible decay during storage.

A synthetic plant growth regulator, 1-Methylcyclopropene, is often also applied to apples after harvesting. This slows down the ripening and maintains the quality of the fruit during extended cold storage and supermarket shelf life.

Storage

Apples can be packed directly from harvest without prior storage, but are first cooled to below 5°C (41°F), and must then be packed within two hours.

All cultivars, except for Granny Smith, must be cooled to -0.5°C (23°F) within five days if they are to be stored, and maintained at this temperature until they are packed.

Early harvested fruits are packed first, followed by post optimum harvested fruit. Apples picked during the optimum harvest window have the highest storage potential and so are maintained either under regular air (RA) or in a controlled atmosphere (CA), with carefully regulated levels of oxygen and carbon dioxide, for packing later. Optimum gas levels vary depending on the variety. Golden Delicious and Red Delicious may be stored for up to nine months under CA, and Granny Smith, Royal Gala and Pink Lady apples ten, seven and six months respectively.

Packing

Apples are graded before packing. Most commercial growers reserve the best quality fruit for export. Fruit that does not make the export grade is either sold on the local market or sent for processing. Apples are usually packed close together to prevent bruising and rub marks. Some pack houses apply an edible coating to the fruit to improve storage quality and extend the shelf life of the fruit.

Inspection

Packed and palleted fruit is checked prior to export to ensure that the quality complies with set standards. Quality checks may include flesh firmness, skin colour, external disorders (including bitter pit), rub marks and bruising, and fruit temperature. The monitoring of pests and diseases, is especially important.

Cooling and export

After inspection, fruit must be forced-air cooled to -0.5°C (23°F) within 72 hours if unbagged or within 96 hours if bagged. Once the correct temperature has been reached, the apples are placed in cooling tunnels to await transportation and distribution.

Below: The highly commercial Braeburn produces attractively marked fruits.

APPLE CULTIVATION AND PLANTING

If apple trees are to do well – growing vigorously and producing abundant crops – they need to be grown in the conditions that best suit them. Take care over planting and new trees will grow away strongly, and you will soon be picking your very own home-grown fruit from them.

Before planting an apple tree it is important to find the right spot to plant it so that it becomes successfully established. Apple trees prefer a position in an open site, although sheltered from strong winds, and in full sun. They will, however, tolerate some shade. You should try to avoid planting trees in positions that may become waterlogged, such as a low-lying part of a predominantly damp garden. Also avoid potential frost pockets, such as low-lying areas where cold air collects.

Soil

Apple trees are tolerant of most soil conditions, but do best in fertile, well-drained soil. They do not do well in very light soils or soils that are prone to waterlogging. However, they tend to grow faster on clay-based soils than on light sandy or chalky soils. The ideal is a crumbly soil with medium fertility and slightly on the acid side. It is not necessary to test the pH level (degree of acidity/alkalinity) of soil when growing apples.

More important is soil structure. To test the soil structure, scrape up a handful of soil, then try rubbing it through your fingers as if making pastry. If it binds into moist crumbs, you have the ideal – a free-draining but fertile loam. If it runs through your fingers easily without binding, the fertility is likely to be low. If you can squeeze it into a solid lump in your hands, you have a heavy clay that will be wet and cold but potentially high in nutrients.

Growing apples organically

It is possible to grow apples successfully without the use of chemicals as fertilizer and to control pests and diseases.

- Choose disease-resistant varieties
- Prepare the ground well before planting, incorporating plenty of organic matter into the soil.
- Clear away all fallen leaves and fruits in autumn that could harbour disease.
- Prune the canopy to improve air circulation among the branches.
- Thin the fruits to two per spur.
- If it is necessary to spray trees, use an organic product.

Soil improvement

Adding organic matter to the soil fulfils two functions. It improves the structure of all soil types, binding thin, sandy soils into larger crumbs and opening up heavy clay, allowing water (and nutrients) to pass through freely. Organic matter also raises fertility, promoting vigorous growth and good disease-resistance.

If you have a heavy clay soil, digging in a bucketful of horticultural grit before planting will improve the drainage. The following materials can be used to improve the soil.

Garden compost

The best of all soil improvers is garden compost that is made in a compost heap or bin. The best compost is made from a mix of organic materials: raw vegetable peelings, eggshells, used tea bags, coffee grounds, annual weeds and other leafy material from the garden (including lawn clippings), newspaper, cardboard and old cotton rags.

Leaf mould

Gather up the leaves from deciduous trees in autumn and stack them in black plastic bags. Tie up the bags then pierce a few holes in the sides to allow for air circulation. Store the bags in an out-of-the-way corner. It can take up to two years for the leaves to disintegrate into a crumbly material suitable for garden use.

Above: Create your own compost from organic household and garden waste.

Animal manures

The manure of vegetarian farmyard animals (cows, sheep and horses) can be used as a soil improver provided it is stacked first and allowed to rot down for three to six months – or it can be added to a compost heap as an accelerator. (Note that animal manures often contain weed seeds.)

Bird manures

The manure of chickens and pigeons is often used in gardens, but care should be taken – it is very high in uric acid, which can 'scorch' plants. The material is best treated as an activator and used as an addition to a compost heap.

Spent mushroom compost

Compost used for growing mushrooms is sometimes available. It contains chalk, an alkaline material, so should not be used on soils that are already very alkaline.

Soil improvers

These are sold bagged up like potting compost and are suitable for use by gardeners who are unable to make their own compost. They have the advantage of being weed free.

Buying apple trees

Apple trees are sold usually as either one- or two-year-old plants. A two-year-old tree will produce fruit sooner after planting than a one-year-old. Young trees without lateral branches are often referred to as whips.

Trees are available container-grown or as bare-root plants. Container-grown plants, which are more expensive, are available year-round. Bare-root trees are sold only during the dormant period when they are out of leaf – in late autumn and winter.

Choice of variety depends on several factors. You should consider not only the flavour of the apple, but its pollination requirements, yield and keeping qualities. Some varieties are hardier than others. It is best to buy from a local nursery, who can advise on which varieties will do well in your soil type and on your site. Modern varieties often have higher levels of disease resistance than older ones.

Rootstocks

All apple trees that are sold in nurseries and garden centres are made up of two separate plants that have been grafted together, or attached until they grow together as one plant. The part of the tree that will form the main trunk and branches is grown from a cutting (the 'scion') that is grafted on to the roots and lower trunk of another (the 'rootstock'). The graft union is clearly identifiable as visible scarring around 25cm (10in) above ground level on a young tree. Rootstocks influence the growth rate and ultimate size of the tree and can also promote disease-resistance. The most widely available are:

- M27 producing a tree 1.5–1.8m (5–6ft) tall
- M9 producing a tree 2.4–3.6m (8–12ft) tall
- M106 producing a tree 3.6–5.4m (12–18ft) tall

Other rootstocks include Budagovsky 9, M26, Mark, Ottawa 3, M7, M2, M4 and MM111.

East Malling Research Station

A research station was established at East Malling in Kent, England, in 1913 on the impetus of local fruit growers. An important part of the station's work has been the development and testing of rootstocks. These are now standardized as the 'Malling Series', identifiable through the initial M, as in 'M27'.

Research into and testing of rootstocks is ongoing. The aim is to produce rootstocks that promote resistance to common apple problems such as fireblight, woolly aphid and collar (or crown) rot as well as improved frost resistance, vigour and high yield efficiency (large fruits in quantity).

Advantages of autumn planting

All hardy trees and shrubs experience a surge in root growth during autumn, even though growth above ground has stopped and the plant appears dormant. An apple tree planted in autumn can thus establish itself in the ground before putting on any new top growth. Growth above ground the following spring will be vigorous.

Family trees

An apple tree that comprises two or three separate varieties grafted on to a single rootstock is called a family tree. They are a good choice if you have room in your garden for only one tree. The varieties on any one tree are all compatible, so the tree is self-fertile and planting a pollinator is unnecessary. However, each variety usually ripens its fruits at different times, thus extending the season.

Family trees are usually grafted on to a dwarfing rootstock and reach around 3m (10ft) in height when mature. These trees can be of ungainly appearance, however, due to the variation in growth rates among the different cultivars.

Right: *Apples are not only delicious when they are picked straight from the tree, but they also retain their qualities when stored.*

Above: *A bare-rooted tree should be planted as soon as possible.*

Sourcing unusual varieties

Even larger nurseries carry only a limited range of apple trees – popular varieties that are likely to do well in the majority of situations.

If you wish to grow an old or unusual variety, you may need to contact a grower or orchardist who owns a tree and ask them to graft it for you. Grafting can take place only at certain times of year, and there needs to be a time allowance for the graft to take. Therefore, you may have to wait for up to a year or so before you are able to bring the tree home ready to plant in your garden.

Planting a bare-root apple tree

1 Before planting, place the tree in a large plastic bucket. Fill with water to cover the root system. Allow the roots to soak for at least an hour.

2 Dig over the soil and remove any large stones and all traces of weeds. Incorporate organic matter to improve the soil, if necessary.

3 Excavate a deep hole that is large enough to contain the tree's root system comfortably. Loosen the soil at the bottom of the hole.

4 Lay a cane across the hole to check the planting depth. The tree must be planted to the same depth as it was in the nursery. Look for the soil mark near the base of the stem for guidance.

5 To ensure the tree grows straight, use a stake to support it. Insert the stake into the ground next to the trunk. Be careful not to damage the root system during the process.

6 Begin to backfill the hole with the excavated soil. As you do so, gently shake the tree periodically to help settle the soil around the roots.

7 Firm the soil around the base of the tree with your hands or your foot. Do not press too hard, or you may compact the surface.

8 Tie the trunk to the stake using a rubber tree tie. Tighten the tie, but make sure it is loose enough around the stem to allow this to thicken.

9 Water the tree well with a watering can fitted with a fine rose. Note that water delivered as a jet from a hose can compact the soil surface.

Planting

Bare-root trees should be planted as soon after purchase as possible. But you may need to delay planting if the ground is frozen or waterlogged. You can store bare-root plants unopened for up to four weeks in a cool but sheltered place, for instance in a garage or shed or other outbuilding.

You should also plant out container-grown trees as soon as is practical, but if this is not possible, stand them in a sheltered but light spot in the garden. Water them frequently if the weather is dry to prevent the compost from drying out. Alternatively, dig a hole in a spare piece of ground and sink the container into it to keep the roots cool.

Heeling in

If you are unable to plant a bare-root plant for several weeks, it is worth considering 'heeling in'.

Dig a shallow trench about 30cm (12in) deep in good soil in a sheltered part of the garden, for instance in a bare vegetable plot. Lay the roots in the trench (angling them if necessary) then lightly cover them with the excavated soil. This will keep the roots cool and moist until you plant the tree.

Planting a container-grown tree

Before planting, prepare the soil as for a bare-root tree. Dig a large hole at least twice the depth and width of the container the tree is already in. Fork in organic matter at the base of the hole.

Below: Tree guards help to protect the bark of young trees until it thickens.

Slide the root ball from the container (it can be beneficial to water the container first, if the compost is dry – it helps consolidate the compost around the roots). Place the tree in the hole. Check the planting depth by laying a cane across the hole. The top surface of the compost should be level with the surrounding soil. (Plant too deep and the base of the trunk can rot where it is in contact with the soil. Plant too shallow and the roots will be exposed to light and air.)

Backfill by sprinkling the excavated soil around the root ball. Lightly firm the plant in with your foot. Water the plant well, using a watering can.

Staking

To protect the newly planted tree from wind-rock, which could disturb the roots, use a short stake. This allows the upper portion of the tree to blow in the wind, which strengthens the trunk in the long term.

Above: An example of a family tree, with Jonagold (left) and Elstar Gray (right) growing from the same trunk.

Insert the stake as you plant the tree rather than afterwards to be sure of avoiding damage to the roots. When staking a tree, either drive the stake in vertically next to the trunk or at an angle into the prevailing wind. Tie the tree to the stake with a special rubber tree tie. Do not pull this tight around the trunk, but leave a gap to allow the stem to thicken. Loosen the tie as the tree grows. Once the tree is well-established (after two to three years), the stake can be removed.

Tree guards

To prevent damage from mice and rabbits, protect the lower portion of the trunk with a tree guard – a length of plastic (or other synthetic material) that is formed either as a strip that winds round the trunk, which expands as the trunk thickens, or as a loose-fitting sleeve. Once the tree has developed thick bark, the guard can be removed.

GROWING APPLE TREES IN CONTAINERS

Apples can be grown successfully in containers, but they need more care and attention than trees grown in the open garden or in a kitchen garden. It's essential to choose a tree on a dwarfing rootstock. Look for a container-grown tree with evenly spaced top-growth (or crown).

Apple trees should be grown in large containers, ideally 40cm (16in) across and deep or even bigger. When in fruit, the tree will be heavy, so to provide ballast and to prevent it from blowing over, you should choose a container made of a heavy material, such as terracotta or reconstituted stone. If you prefer to use plastic, resin or some other lightweight synthetic material, place large stones or bricks at the base before filling with compost to create stability.

Even so, apple trees in containers should be positioned so that they are sheltered from strong winds. Choose weaker-growing varieties that have been grafted on to dwarfing rootstocks to limit their growth. Spur-bearing types will produce the most fruit.

Below: Choose a tree on a dwarfing rootstock for growing in a container and keep the tree well watered throughout the growing season.

Above: A balcony garden with Malus *Klarapfel and Cox Orange growing in pots.*

Above: Containers should be positioned where they will receive plenty of sunshine.

Compost

For the best results, use a soil-based compost. John Innes No.3, formulated for trees and shrubs, contains the appropriate levels of nutrients.

Multi-purpose composts can also be used for apples in containers, but you will need to pay more attention to watering, as they are difficult to wet if allowed to dry out.

Garden compost is unsuitable for use in containers. It is not sterile and contains too many micro-organisms and invertebrates. However, if you have a source of leaf mould, this can be added to proprietary compost.

Watering and feeding

Water apple trees in containers regularly, especially during dry weather in spring and summer. It is very important to maintain water levels while the fruit is swelling in summer. Applying a potassium-high fertilizer, such as a rose fertilizer, in spring will boost flower and fruit production. A further dose in autumn helps firm the growth before winter. Alternatively, apply a tomato fertilizer as a root drench when the tree is in vigorous growth in spring and early summer, and again in autumn.

In winter, water just enough to prevent the compost from drying out completely. The aim is to keep the roots just moist. Do not water during freezing weather.

General care

Apple trees in containers will need pollinators just the same as trees in the open garden. If there are no other apple trees in your own garden or one nearby, grow several different ones together or choose a self-fertile variety or family apple tree.

Watch out for pests and diseases and deal with them as soon as you detect any sign of damage. Prune the trees carefully – cutting the stems too hard will result in overly vigorous growth that will not bear fruit successfully the following year.

Although many apple trees can be grown in containers, they should not be viewed as long-term plantings. They are long-lived trees and sooner or later any tree in a container will need planting out in the open if it is to thrive.

Winter care

All plants in containers are susceptible to cold in winter, as the roots are above ground level and liable to freeze. If a hard frost is forecast overnight, either move plants under cover – for instance, into a porch, unheated greenhouse or shed – or wrap a length of horticultural fleece around the container to insulate the roots.

As apples flower relatively late in the season, it is not usually necessary to protect the blossom from frost. But if a hard frost is forecast when the tree is in flower, lightly tent it with horticultural fleece. Remove the fleece during the daytime. You can leave the fleece in position if the temperature remains below freezing during the hours of daylight.

Repotting

When the tree's roots fill the container, it is necessary to repot it.

Tilt the container on its side and carefully slide the plant out. Shake the roots free of compost, washing them under an outside tap to dislodge any that clings on. Scrub the interior of the container. Lightly prune the roots, then return the tree to the container, using fresh compost. If this is not practical, tilt the container and scrape out the uppermost layer of compost with your fingers. Replace this with fresh compost or leaf mould.

Buying an apple tree for a container

It is a myth that you can keep an apple tree small by pruning the roots. You must buy a tree on a dwarfing rootstock. The best one for growing in a container is the Malling rootstock M27.

Planting in a container

1 Line the base of the container with crocks or stones, making sure all holes at the base of the container are covered. This will allow excess water to drain freely while preventing loss of compost through the holes.

2 Begin to fill the container with compost or leaf mould, to a depth of around one-third. Slide the plant from its container. Place it on top of the compost, in the middle of the prepared container.

3 Check the level by placing a cane or straight length of wood across the container. The top of the rootball should be about 2.5cm (1in) below the rim of the container, to allow for watering and maintenance.

4 Fill the gap between the edge of the container and the rootball with more compost. Press fertilizer pellets into the compost, midway between the rim of the container and the rootball.

5 Water the container well using a can fitted with a fine rose or from a hose with the nozzle set to produce a fine spray. This avoids compacting the compost surface.

6 Top-dress the compost surface with small stones to keep the roots cool and prevent excess moisture loss through evaporation. It also makes the pot look tidy and attractive.

TRAINING APPLE TREES

The typical image of a tree is usually a large plant with a solid trunk and a rounded system of branches, but there are also a wide variety of forms into which trees can be trained. Apple trees are trained to produce compact plants that flower and fruit prolifically, and present the crop around eye level.

Most fruit trees can be bought ready trained and this is the most practical option for busy gardeners, but it can be rewarding – and is much cheaper – to train your own. Ready-trained trees will flower and fruit the first season after planting, but if you are training the plant yourself, no crops will be produced until, usually, the third year.

Size

There are several reasons for choosing a particular shape of fruit tree. The first is size. A fully grown apple tree, for example, can take up a lot of space, especially in a small garden, but a dwarf pyramid can be fitted into a limited space, while cordons can be grown along a fence or even as a hedge.

Productivity and quality

Training fruit trees helps to improve productivity and quality. The upper branches of a standard tree can shade the lower branches, and these in turn will shade the ground beneath it, limiting what you can grow there. However, a fan grown against a wall will not only produce a large crop but will supply individual fruits with the maximum amount of light for even ripening. Training trees against walls also provides protection and warmth.

Above: In the kitchen garden apple trees can be grown as vertical cordons to save on space.

Central leader tree

This form eventually produces a large tree, suitable for use as a specimen.

Allow a young tree or whip to develop with minimal pruning. Once the tree is established – after three to five years – begin to remove the lower branches to create a clear trunk 1.5–2m (5–6ft) tall. The main upright stem, or the leader, is allowed to continue growing so that the tree achieves its natural dimensions.

If growing apple trees primarily for their fruit, this training method is not recommended. Crops will be small and difficult to harvest as many of the fruits will be borne well above head height.

More varieties

When choosing the shape of your tree you should also think about the number of varieties you require. For example, a fan might occupy the whole of one wall, but in the same amount of space you could grow half a dozen cordons or more. These will not necessarily yield more fruit but could provide variety of flavour and availability.

Decorative qualities

Do not overlook the decorative aspect of trained trees. A tall espalier growing against the end of a house can be stunning, as can fans on a smaller scale. Free-standing cordons and espaliers produce excellent screens.

Choosing trees

Before buying a tree consider the above aspects and think about where you are going to plant it and what the best shape is for the space available. It is also worth remembering that basic shapes, such as standards, require far less pruning than more complicated ones. Something to bear in mind, especially if you are not happy on ladders or steps, is that the fruit on larger forms may be out of reach.

The following are some of the fruit tree forms that are most commonly used by amateur gardeners. Basic definitions of the forms are given here, with details on how to create them over several years.

Standard/half-standard

Standard trees are grown on vigorous rootstocks and are usually too large for most gardens. They have a clear trunk of 2–2.1m (6–6½ ft) – 1.35m (4ft) for a half-standard.

To develop the standard (or half-standard), allow a young tree to grow unpruned, clearing the trunk of laterals as for a central leader tree. Once the main stem has achieved the desired height, cut it back, then prune the laterals as for a bush.

Bush

This is an excellent form for any garden where a number of fruit trees are grown, but they must be on dwarfing rootstocks. The aim is to create a bushy plant on a trunk 75–90cm (2½–3ft) tall.

Prune the leader of a young tree or whip to stimulate branching low down. In winter, select the strongest laterals to form the framework and cut back the remainder. Shorten the remaining laterals to outward-facing buds to develop the desired open centre. The following winter, lightly prune these to encourage bushiness.

Above: Training trees as cordons is the perfect way to get several trees into one small garden.

Pyramid

A pyramid is a dainty tree with a strong central leader. Staking is essential, as this tree form is less stable than a bush. The first winter after planting, shorten the leader to 50–75cm (20–30in). Develop lower branches first, cutting back suitably placed laterals to outward-facing buds on the undersides. This encourages a horizontal habit. It is important to remove vigorous shoots on the upper portions of the leader.

As the plant grows, tie in a replacement leader to the stake to develop height, removing all other vigorous upright-growing branches. Prune the upper branches in the same way as the strong laterals towards the base of the plant, cutting to outward-facing buds on the undersides.

Cordon

A cordon is a vigorous upright that has usually been trained at an angle, with stubby side shoots – ideal where space is limited. The aim is to build up a system of short, stubby 'spurs' the length of the stem.

A cordon must be grown against a system of wires. The main stem is tied to a sturdy cane, and this in turn is tied to the wires. To maintain the cordon, remove overcrowded growth in winter and shorten overlong laterals to three or four buds. Shorten the leader, as necessary, in late spring. In summer, cut back any overlong, whippy shoots to the base. A double cordon is similar to a cordon but is upright and has two main stems. Double cordons are obviously more productive than single ones.

Espalier

In this form, lateral branches are trained strictly horizontally, either against a wall or freestanding on wires.

Unless the tree is bought ready trained, plant a whip and cut it back in winter to a strong bud just above the lowest wire. From the new growth, select the strongest as the new leader and tie this to an upright cane lashed to the wires. Select two strong laterals and tie these to canes tied in diagonally to the wires. The next winter, bring these down to the horizontal and attach them to the lowest wire. Cut back the leader just above the second wire and continue this procedure in subsequent years until the espalier is complete.

Fan

A decorative form, also trained on wires, that is useful for varieties that benefit from wall protection. Training is less strict than for an espalier.

Unless bought ready trained, choose a young plant with several strong laterals. On planting, cut all of these back apart from two that are suitably placed for training to the diagonal to either side of the leader. Tie them to canes that can be lashed to the wires. Cut back the leader to just above the upper 'arm'. As suitably placed strong side shoots grow, tie them in to the wires, aiming for an even development to both sides.

Above: The espaliered tree is decorative and highly productive, capable of producing many tiers of fruit-bearing branches.

Stepover

This is an ideal form to use to edge beds in a kitchen garden. Stepovers are suitable only for apples on a dwarfing rootstock.

Cut back a whip to within 30cm (12in) of the ground. Pull the new leader hard down to the horizontal and train it on a horizontal wire stretched between short uprights.

Below: This stepover Malus *'Red Devil' has been grown on M25 dwarfing rootstock.*

PRUNING APPLE TREES

Many gardeners are confused about pruning, but it is not really difficult – the main thing is knowing when to do it, and how much to remove. While failing to prune at all may not cause serious damage, it will not make for the most healthy and productive trees, and they may also become too large.

Apple trees need a regular pruning regime if they are to achieve their optimum cropping. They bear fruit on stems that are in their second year or older. Shoots produced in the current year will flower and fruit in the following year. When pruning, consider the tree's future cropping and aim to produce an equal balance between older shoots and new ones.

When to prune

Apple trees are pruned in mid- to late winter. Trained trees, once established, should also be pruned in summer. Winter pruning invigorates a plant, encouraging it to grow more during the coming season. Summer pruning can inhibit growth, as you are removing a proportion of that year's growth. It helps maintain the form of cordons and espaliers.

Apples trained formally as fans, cordons and espaliers need to be pruned in summer as well as winter. If they are left unpruned, new stems will grow vigorously. Not only will these spoil the tree's form, but they will be relatively unproductive and will cast unwanted shade over fruits developing lower down the stem. The method used for summer pruning of trained apples is known as the Lorette system. In late summer, when the lower third of the new growth is starting to firm up, shorten all new shoots. On shoots arising directly from the main stem, count three or five leaves above the basal cluster of leaves, then cut just above the leaf joint. Cut shoots that have grown from existing side shoots back just above one leaf above the basal cluster of leaves.

Above: A pole or pillar apple takes up very little space but can be very productive. It is pruned as for cordons.

Above: Cut back all the new, straight growth, which, if allowed to develop, will make crowding and unproductivity worse.

Above: Reduce the number of spurs by pruning a number of the older ones. Leave enough to produce the next autumn's fruit.

Summer pruning carries a certain risk as cutting stems encourages new growth, which can be vulnerable to frost damage. Shorten any long, whippy shoots that appear to a growth bud near the base in mid-autumn.

Pruning cuts

Although different fruit trees and styles of training involve different pruning methods, the pruning cuts are the same in all instances. Use sharp secateurs (pruners), loppers or a pruning knife and make sure you are comfortably positioned to make a clean, neat cut. If you need a ladder, either make sure that someone will hold it for you or use one with an integral stabilizer.

Always cut a stem just above a bud and make sure the cut is angled away from the bud. Do not cut too far above the bud, or the stem will die back, possibly allowing disease to enter the tissue. Cutting too close to the bud may damage it, preventing it from growing and encouraging disease.

Branches that are large enough to be cut with a saw are usually cut across at right angles. If the branch is heavy and and likely to split, the sawing is usually done in three stages (see opposite).

Shortening leaders

The leading shoots, or branch leaders, are the main shoots of a tree. You should shorten the previous year's extension growth by up to one-third. Very vigorous shoots on spur-bearers are best left unpruned, as pruning will stimulate even more vigorous growth.

Spur- and tip-bearers

Apples are described as being tip-bearers or spur-bearers, depending on how they carry their fruit. Tip-bearers produce their fruits at or near the tips of branches, while spur-bearers produce their fruits along the branches on short, stubby growths known as spurs.

The 'June drop'
Apple trees regularly shed a proportion of their fruits in late spring. This is a natural phenomenon and does not indicate that anything is wrong. It is common for a tree to set more fruits than can develop fully, and shedding the excess is a means of ensuring that the remainder will ripen correctly.

Most varieties are spur-bearing, but among these a few are partially tip-bearing. Tip-bearers produce smaller crops than spur-bearers, and should be grown only as standard or half-standard trees or as bushes. Spur-bearers lend themselves to training.

Pruning a spur-bearing apple

The aim in pruning a spur-bearing apple is to produce and maintain the short, stubby growths that will flower and bear fruit.

Shortening side shoots causes them to thicken and produce clusters of flowering shoots. Side shoots fan out from the tree's main shoots. In mid- to late winter, cut these back to five or six growth buds from the base of the shoot. Weak-growing shoots can be pruned harder. Shoots that are already short can be left unpruned.

After around five years, spurs tend to become congested and therefore less productive – they produce too many fruits too close together that cannot grow to full size or ripen evenly.

On each congested spur system, cut out the lower spurs at the base. Thin the remaining upper spurs to leave about three well-spaced spurs. After thinning, each spur system should be capable of bearing four or five good-sized fruits.

Pruning a tip-bearing apple

The aim in pruning a tip-bearing apple is to relieve congestion in the body of the tree and to remove older growth in favour of new wood that will bear future crops.

Shorten old stems that have fruited in previous years. Cut them back either to new shoots that grew during the previous summer (these will flower and fruit during the coming year) or to strong growth buds lower down on the stem (these will shoot during the coming year then fruit and flower the next). You do not need to remove all the old wood in any one year but can leave some stems unpruned. The general aim is to encourage the tree to replace its fruiting wood over a period of years.

Pruning a large tree

Many gardens are home to a large, mature apple tree that is valued as much for its ornamental appearance and home for wildlife as it is as a fruit-bearer. While a systematic pruning regime is not practical, it can be beneficial from time to time to remove older and crossing branches to open up the heart of the tree and rejuvenate it. If you need to reduce the size of a tree drastically (for instance, if it is overhanging a neighbour's garden or is too close to the house), consult a qualified tree surgeon.

Larger branches on mature trees can sometimes be torn off by strong winds, or even by the weight of the crop. It can also sometimes be necessary to remove a large branch to restore the symmetry and balance of a tree, especially on large specimens that have not been strictly pruned. To prevent further damage to the tree, cut away the branch in stages. It is best to carry out this type of pruning in winter, when the tree is dormant.

Below: Carefully cut back congested or crossing branches to the point of origin.

Above: When pruning a cordon, any new shoots on existing side shoots should be cut back to one leaf.

The first cut is made on the underside of the branch, 5cm (2in) out from the final cutting position. The second cut is made slightly further out along the branch, this time from above, by sawing down until the branch splits along to the first cut and is then severed. The final cut can be made straight through from the top because there is no weight to cause splitting.

Fruit thinning

If a tree has set a large amount of fruit during spring, it can be beneficial to thin the fruits out in early summer. This results in fewer apples, but larger and of better quality, and a greater total weight, and helps offset a potential tendency to biennial bearing. Large fruit clusters will also weigh down branches, which can sometimes snap if they are brittle. A further aim of fruit thinning is to expose the developing fruits to the sun, so that they ripen fully and evenly with good skin colour.

The time to thin the fruits is after the 'June drop'. Remove the central fruit from each cluster. Cut out any other fruits that are damaged or mis-shapen. Around two to four weeks later, thin again to leave one fruit per cluster.

PROPAGATING APPLES

Apples are usually propagated by means of grafting. Cuttings taken in the usual way will not produce good plants. The techniques generally used have been developed over the course of many centuries and have proved to be extremely reliable.

It is not possible to increase stocks of named varieties by seed – but seed-raised plants can make attractive ornamentals and may act as pollinators to other apple trees.

Growing from seed

Place a handful of perlite in a clear plastic bag, moisten slightly, add some apple pips and refrigerate. After about six weeks, remove the seeds from the perlite. Fill small containers with multi-purpose or seed compost, water and allow to drain, then press the seeds into the compost surface. Top with a thin layer of horticultural grit or sharp sand. Place the containers in a sheltered space outdoors or in a cool greenhouse.

Once the seed has germinated, keep the seedlings growing strongly by making sure that you water them regularly. Water less from late autumn and water in winter only to prevent the compost from drying out. The following spring, the seedlings should be large enough to pot up individually, using a soil-based compost.

Feed the young plants with a dilute tomato feed. Transfer them to larger pots once the roots fill the existing ones. The young trees should be large enough to plant out after two to three years and should flower and bear decorative fruits from five years.

Grafting

Widely used in the nursery trade as a means of propagating many trees and shrubs, grafting allows new plants of saleable size to be produced quickly from just a small amount of plant material. Fruit trees are nearly always grafted.

Grafting is a technique that involves uniting a cutting or growth bud (known as the scion) with the lower parts (the rootstock) of another. The following techniques are the ones usually used for apples.

Above: Malus *scions, successfully grafted under the bark of an old apple tree trunk.*

Preparation of the rootstocks

Rootstocks should be well-established at the time of grafting. They are usually sold as dormant plants and should be planted as described for planting a bare-root tree (*see* p.18). They can then be used for chip-budding the following summer and for stem grafting the following late winter.

After planting, remove any side shoots from the lower 30cm (12in) of trunk. Rub out any buds that appear on this part of the rootstock during the growing season. Keep the rootstocks well-watered during spring and summer. This enables you to lift the bark easily when chip-budding.

Below: *The sequence of chip budding.*

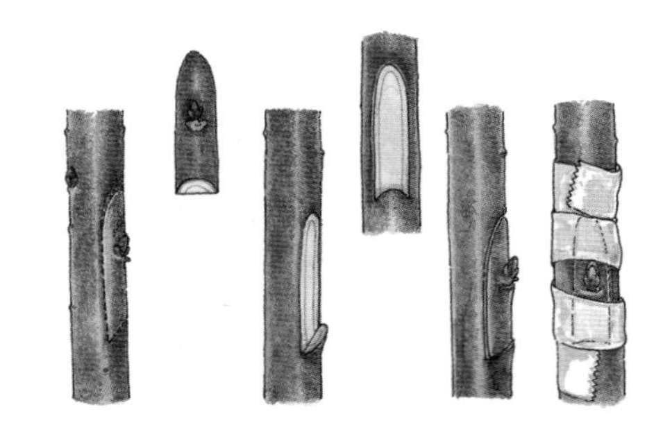

Principles of grafting

For a successful graft, there must be a free flow of sap between the rootstock and scion and the tissues of the two elements should knit together.

Below the bark on a stem is a layer of tissue called the cambium layer which must be aligned so that the stock and scion can fuse together.

Bud-grafting

This method is popular with nurserymen as it is possible to produce a new tree from a single growth bud – hence several new plants can be created from a single stem from the variety you wish to propagate.

Chip-budding

This form of grafting is carried out in mid- to late summer.

Cut well-ripened stems of around pencil thickness from the sunny side of the tree. To prevent moisture loss, strip off the leaves, keeping the leaf stalks intact. Remove any small new leaves that appear at the base of each leaf stalk. To prevent further drying out, wrap the stems in a damp cloth.

Below a healthy, well-developed bud (about 2cm/¾in), make a straight, shallow, downward-angled cut. Starting about 2cm (¾in) above the bud, make a downward cut, cutting behind the bud and down to meet the inner edge of the first cut, removing a 4cm (1½in) sliver of wood.

On the rootstock, locate an area of stem about 23cm (9in) above ground level that is free from buds. Remove a section of wood corresponding in shape and size to the bud. Fit the bud into the rootstock. It should fit exactly, but if the bud is slimmer than the rootstock, line up the bark to one side so that the cambium layers of the scion and rootstock are in contact.

Tie the bud in place with grafting tape. Tie loosely around the bud itself, to allow room for this to grow, but sufficiently tightly to hold it in place.

Aftercare of chip-budding

After about four to six weeks, the wound should start to callus over as the graft takes. Remove the tape. If the graft has been successful, the petiole will be shed in autumn at the same time as the rootstock loses its leaves. The following spring, cut back the rootstock just above the grafted bud. To encourage upright growth, tie the stem to an upright cane inserted into the ground next to the rootstock. For maximum growth in this early stage, remove any flowers that form. Cut back any shoots that appear on the rootstock, but only when they have reached a length of 5–7.5cm (2–3in). The tree can be planted out in its final position in the autumn.

Whip-and-tongue grafting

This grafting method is carried out in late winter to early spring around the time that sap is starting to rise.

The scions are prepared in mid-winter while the tree is fully dormant. Take cuttings of the previous season's growth about 23cm (9in) long. Cut just above a bud on the parent plant.

These cuttings then need to be kept cool and in a state of dormancy until they are grafted a few weeks later. Taking the cuttings in advance means that when you come to do the graft all the fresh sap will be coming from the rootstock. Bundle the cuttings loosely together and bury them to two-thirds of their length in a sheltered part of the garden. Alternatively, seal them in a plastic bag and refrigerate.

When the rootstock starts into growth, cut it back, about 23cm (9in) above ground level, with a downward-sloping cut. Remove any side shoots below the cut. With the cut sloping down away from you, make a second upward cut around 3.5cm (1½in) long to remove a sliver of bark and expose the cambium layer and pith.

About one-third of the way down this second cut, remove a narrow V-shaped section of wood to form a 'tongue', into which you will later insert the scion.

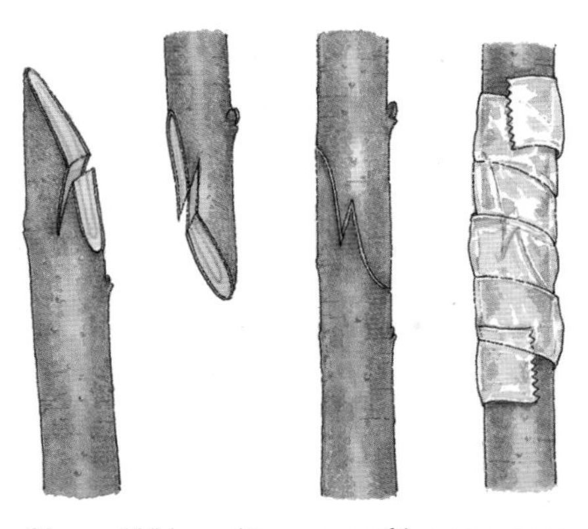

Above: *Whip-and-tongue grafting sequence.*

To prepare the scion, first trim any soft growth from the tip. Then shorten it at the base to a length of wood that holds three or four buds. Trim the base of the cutting about 3.5cm (1½in) below the lowest bud.

From the base of the cutting, on the same side as the lowest bud, make a slanting cut through to the other side, the same length as the cut on the rootstock. At a point on the cut surface equivalent to the 'tongue' on the rootstock, make a short upward cut. Gently turn the knife to open this, then insert the scion into the rootstock.

Bind the union with grafting tape. The graft should 'take' within four to six weeks, after which the tape can be carefully removed.

Note: When inserting the scion, the cambium layers must line up. If the scion is thinner than the rootstock, place it so that the cambium and outer bark line up to one side.

Aftercare of whip-and-tongue grafting

As the buds on the scion begin to shoot, select the strongest to form the leader of the tree. Shorten the remainder to about 7cm (3in). To encourage upright growth, insert a cane into the ground next to the rootstock and tie the selected shoot to it. Continue to tie it to the cane as it grows. Remove any shoots that appear on the rootstock once they are about 7cm (3in) long. The new tree can be transplanted the following autumn or the spring after that.

Rind grafting

This is a form of grafting over. Instead of using a seed-raised rootstock, scions are grafted on to an established tree, the aim being to test out a new variety or to provide a pollinator compatible with other apple trees growing nearby.

In early spring cut back most of the main branches on the tree to within 60cm (24in) of the main trunk, leaving a few to carry on growing normally. Cut the branches horizontally across.

For the scions, take cuttings of the previous season's growth, each with three growth buds. Trim them just above the uppermost bud with a cut angled away from it. Trim the base of each scion with a sloping cut about 2.5cm (1in) long.

Cutting downwards, make two or three slits in the bark about 2.5cm (1in) long and evenly spaced around each prepared branch of the tree. With the tip of the knife, ease back the bark. Slide each scion into a slit, so that the cut surface at the base of the cutting is in contact with the tree tissue beneath the bark. Bind the union with soft twine and seal with grafting wax.

Remove the binding when the graft has taken and the stems are growing strongly. Cut back the weaker growing stems so that only one remains. It should flower and fruit within three or four years.

It is possible to rind graft several different varieties on to a tree to produce a family apple tree.

Below: *The sequence of rind grafting.*

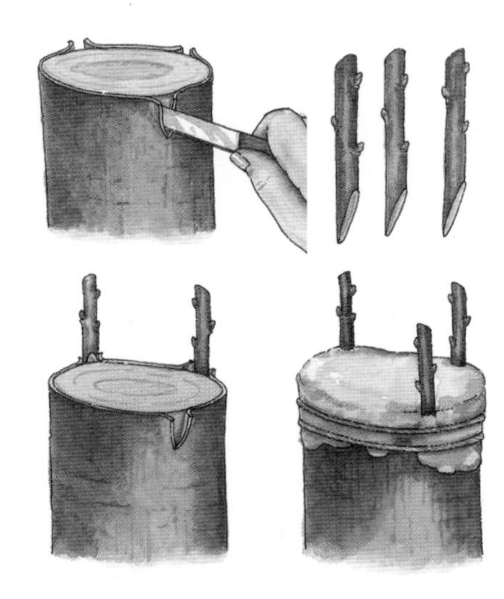

HARVESTING AND STORING APPLES

A principal reason for the enduring popularity of apples is the ease with which the fruit can be stored. Home-grown apples can be enjoyed over a period of many months provided they are harvested carefully, prepared well and stored in appropriate conditions.

Apples are ready to pick when ripe, unblemished fruits start to fall from the tree. The base colour is a clue to ripeness. It should be an even yellow or yellowish-green. On red-skinned varieties, look for this colour around the stem cavity – if it is still mid- or dark green, the fruit is not yet ripe.

Ripe apples should be easy to pick with the stem still attached. Gently roll or twist the apple in your hand so that the stem separates from the branch. The fruits will store best if the stalk is retained.

Over-ripe fruit is not suitable for storing.

Above: Apples should be harvested by hand with a gentle twist of the wrist.

Above: Apples should be well spaced during storage to prevent any rot from spreading.

Storing apples

Apples can be kept for varying lengths of time, depending on the variety. A few seem even to improve during storage, as this allows the flavour to develop fully.

Fruits intended for storage must be unblemished. Ideally, they should be stored at a temperature of 0°C (32°F), to slow down further ripening.

Apples should be individually wrapped so that they do not touch each other while being stored. Otherwise, any rot that develops in an individual fruit will be rapidly passed on to the others.

Place the fruits in open crates or wooden boxes, then keep them in a cool, dry, dark place such as a shed, outbuilding, cellar or garage. Crates for stacking should have slits in the sides or other openings that allow for good air circulation. The storage area should be well ventilated – excessive damp will cause the fruits to rot.

Apples can also be stored in a domestic refrigerator. Fill plastic food bags with firm but ripe, unblemished fruits then seal them. Pierce a few holes in the bags with a wooden skewer to allow free passage of air around the fruits.

Preserving apples

Apples that otherwise do not store well can be preserved by drying or freezing.

Sliced apples can be dried in a low oven for up to eight hours. Allow the apple slices to cool completely before storing them in an airtight container.

Wrapping fruits for storage

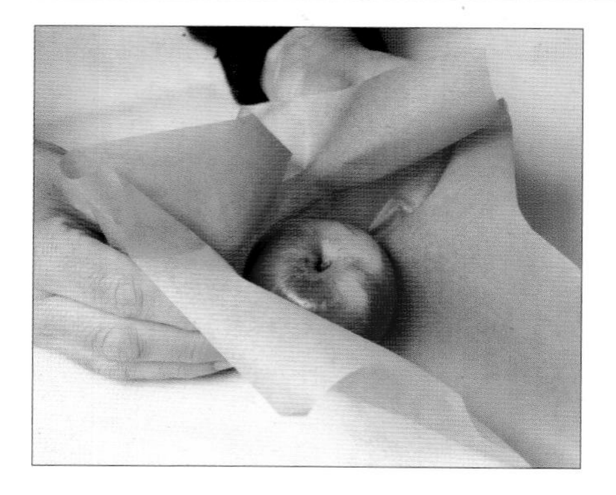

1 Cut pieces of silicone paper or baking parchment, about 30cm (12in) square, or less if the fruits are small. Place a fruit in the centre of each.

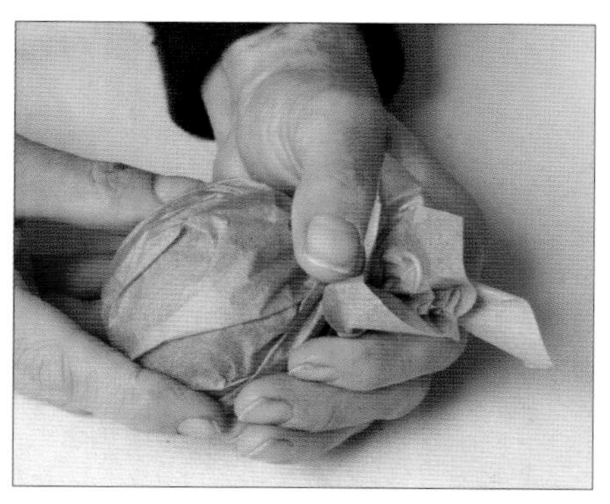

2 Wrap the paper evenly around the fruit, twisting it at the top to secure it.

What happens during storage?

An apple is not dead at the time of harvest and continues to take in oxygen and give off carbon dioxide and ethylene during storage. But since it is no longer attached to the tree, it has to use up food reserves laid down during its period of growth. Consequently, the sugar, starch and acid content of the apple changes. A stored apple can taste sweeter than one eaten straight from the tree. Concurrently, there can also be a loss of firmness. Eventually, the tissues break down and the apple withers and starts to decay.

Apples that store well
Baldwin, Ballarat, Belle de Boskoop, Carlos Queen, Egremont Russet, Freyburg, Goldrush, Granny Smith, Liberty, Lobo, Merton Russet, Pomme d'Api, Primevere, Rhode Island Greening, Splendour, Sunny Brook, Tydeman's Late Orange, Westland

Culinary apples can be cut into pieces and frozen before use. However, they may not hold their shape during cooking as well as unfrozen fruits. Frozen apples are suitable for use in fruit pies and crumbles and for sauce making and juicing.

Juicing and cider making

Extracting juice from an apple is a means of preservation, as the juice is usually pasteurized or distilled as part of the process.

For the maximum nutritional content, apples should be juiced unpeeled. If you are using fruit bought from a greengrocer or supermarket, wash it carefully first to remove the wax with which it will have been sprayed.

Cider

An alcoholic drink, cider is made from the fermented juice of apples. It remains particularly popular in traditional apple growing regions, such as south-west England, Ireland, parts of Spain, Germany and northern France. In North America, the term cider is generally used for non-alcoholic drinks, hard cider being the preferred term for alcoholic ones.

Cider can be made from any variety of apple, but in some regions certain cultivars (referred to as cider apples) are preferred. Many varieties have been bred specifically for cider making. Cider can be sweet, medium or dry and varies in colour from pale yellow to brown. Depending on filtration, it can be clear or cloudy (some apple varieties yield a clear cider without filtration). The finished drink can be either still or sparkling.

Cider can also be made from pears, and is usually called pear cider or perry.

Preparing apples for drying

1 Peel each apple with a sharp knife or vegetable peeler and cut out the core, ideally in one piece, using a corer. Try to avoid damaging the whole fruit.

2 Cut each apple into rings of even thickness, around 6mm (¼in). Place them on a non-stick baking tray and dry in a low oven for several hours.

Preparing apples for freezing

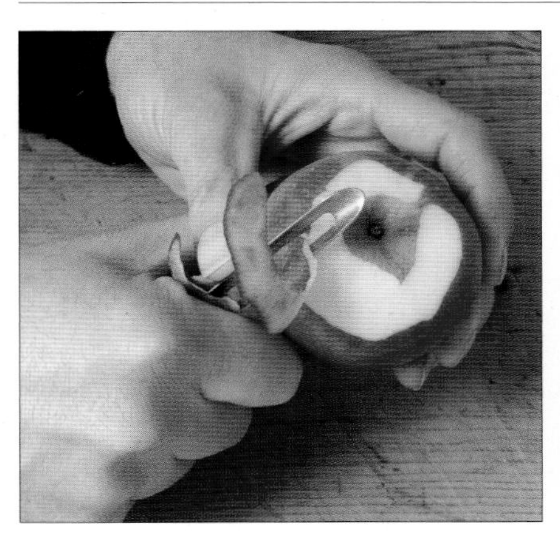

1 Select unblemished apples of even size. Carefully peel each apple with a sharp knife or vegetable peeler.

2 Halve each fruit, then cut into quarters or eighths. Cut out the core and any pips from each segment.

3 Depending on the size of the apple variety, cut each of the segments once more to produce 16 equally sized slices.

4 Bag up the apples in freezer bags, and label each with the date. For the best freezing, suck the air out of the bag before sealing.

APPLE PROBLEMS

Apple trees are subject to a number of problems that can weaken growth, reduce or spoil the crop or occasionally even cause the death of the tree. Problems can be caused by pests and diseases, but sometimes also result from unsuitable growing conditions and sudden adverse changes in the weather.

Identifying a problem is not always easy. Many diseases, for instance, produce symptoms that are similar to those induced by poor growing conditions. However, it is important to deal with any problem as soon as it is noticed. Given appropriate attention, trees usually recover well. It is usually possible to take preventive action so that the problem does not recur.

You can avoid potential difficulties by selecting carefully. Some varieties are known to be resistant to certain diseases, or to perform well in damp soil or in cold areas. A tree that is well-suited to the prevailing conditions will be strong-growing and will recover more quickly than a weak one, should any problem affect it.

Garden hygiene

Practising good garden hygiene can be an important strategy in keeping problems under control. Remove any fallen fruits and leaves from around the tree. They can be a breeding ground for fungi and also shelter invertebrate pests. However, a few windfall apples around a tree will be a valuable food source for birds in winter and beneficial to wasps. Rather than composting these, it is best to burn them. Fungal spores are not killed by the composting process, so you run the risk of returning them to the trees if you use the compost as a mulch around them later on.

Maintaining vigour

Feed and water plants well when they are young to ensure they make rapid progress in the early years. It is also advisable to water established trees during periods of drought in summer, especially if they are bearing heavily. Mulch heavily after watering to conserve soil moisture.

It is often possible to repair any damage to an established tree by judicious pruning.

Attracting wildlife

You can reduce the incidence of pests by attracting as wide a range of wildlife into the garden as possible.

Below: A bat box under the eaves provides shelter for bats during the day.

- Large apple trees will provide a perch and a potential nesting site for birds that will then feed on insect pests. Berrying plants such as hollies (*Ilex*), pyracanthas and cotoneasters will provide a food source in winter. Carefully sited bird boxes (designed for particular species) will encourage roosting.
- Bats will feed on nocturnal moths. Bat boxes in tall trees and under house eaves will give them a shelter during the daytime.
- A small pile of logs or even a tree stump will be home to a range of invertebrates, including stag beetle larvae. Piles of stones will provide a cool resting place for frogs and toads.
- A garden pond will be a habitat and watering hole for many insects, birds and small mammals.

Above: Adding mulch will improve soil structure and fertility as it breaks down.

Garden pests

Pests range from tiny invertebrates (insects and others) to mammals such as rabbits and deer. Some are active at specific times of the year.

Birds

Although they feed on insect pests, birds can be a pest themselves.
DAMAGE: Birds peck at flower buds and ripe fruits. Pecked fruits attract wasps and other undesirable insects.
CONTROL: Deter birds humanely by hanging old CDs or foil trays from the branches that will flash in the light. Change the positions of these regularly. Netting trees is not recommended as small birds can become trapped.

Rabbits

These small burrowing mammals feed on a range of wild and cultivated plants.
DAMAGE: Rabbits can cause serious damage to trees, as they gnaw at the bark at ground level. This exposes the tissue beneath to disease.
CONTROL: Place tree guards around the base of newly planted trees. To keep rabbits out of gardens, erect a boundary fence at least 1.2m (4ft) high. As a guard against burrowing, sink a length of corrugated plastic or metal into the ground along the fence, to a depth of about 45cm (18in).

Above: An otherwise healthy apple has first been pecked by birds and is now attracting wasps.

Deer
Ruminant mammals that are widespread in rural areas.
DAMAGE: Deer browse the branches of trees and bushes.
CONTROL: Erect a fence at least 2m (6ft) high around the perimeter of the garden or plant a tall evergreen hedge to keep them out.

Aphids (greenfly and blackfly)
Sap-sucking insects (Aphidoidea).
SYMPTOMS: The aphids, often clustering on the undersides of leaves, suck the sap, distorting young shoots and leaves. They seldom kill a tree outright, but it may be seriously weakened.
CONTROL: Inspect new growth often for signs of attack, particularly if other related garden plants, such as roses, show signs of damage. Either spray with an insecticide or an insecticidal soap. On a small plant, you can rub off the pests with finger and thumb.

Apple sawfly
Small wasp-like insects (*Hoplocampa testudinea*).
SYMPTOMS: The adult insects first lay their eggs on the blossoms. The eggs then hatch into maggots which tunnel just below the surface of the developing fruit's skin. This causes ribbon-like scars on the outside of the fruit. As the maggots grow, they then tunnel directly into the middle of the fruit, causing it to drop prematurely around mid-summer.
CONTROL: Spray the tree with derris when all the petals have been shed. Remove and burn all affected fruits that show signs of scarring.

James Grieve and Worcester Pearmain are both susceptible varieties.

Winter moth
A moth (*Operophtera brumata*).
SYMPTOMS: Numerous holes are eaten in leaves. This weakens the tree and leaves it susceptible to other problems. Flightless females climb trees in winter, laying their eggs on branches. The emerging caterpillars feed on the leaves, blossoms and young fruitlets during the spring. The caterpillar weaves a silken thread loosely through the leaves and the small holes made at this stage often go unnoticed. As the leaves develop, the holes enlarge and become noticeable. Caterpillars then drop down to the soil where they pupate into adults – they will emerge from the soil anytime during winter.
CONTROL: Tie sticky grease bands around trunks in winter to prevent the females from ascending into the topgrowth. Remove and burn this in spring. (Bands should also be wrapped around any supporting stakes.)

Codling moth
A moth (*Cydia pomonella*).
SYMPTOMS: Maggots are found in fruits. The adult lays its eggs on the surface of the developing fruit in late spring. The larvae then tunnel into the centre. The caterpillar is fully fed around late summer, so it eats its way out of the fruit and spends the winter in loose flakes of bark on the tree trunk.
CONTROL: You can control numbers by hanging sticky pheromone traps in trees. These mimic the scent of sexually mature females. Males are attracted by the scent then are caught in the traps. These traps not only reduce numbers but also indicate the presence of the pest in a garden. Spraying with an insecticide in early summer can control the caterpillars. The pest can also be controlled through the use of a pathogenic nematode (*Steinernema carpocapsae*), which enters the caterpillars and infects them with a bacterial disease. Spray the nematode on the trunk and branches, and also the soil under the branches, in early autumn, after the caterpillars have left the fruit. Note this treatment gives no protection against female codling moths flying in from nearby gardens to lay their eggs the following year.

Woolly aphid
Sucking insects (*Erisoma lanigerum*).
SYMPTOMS: Fluffy white areas appear on bark. The pest is often mistaken for a fungus or mould. The aphid appears in the spring on the bark of some fruit trees – it is common around bark which has not been cleanly pruned. If you rub your finger over them, the aphids will be crushed and wet, which is the proof that the infestation is not mould. The waxy coating makes them difficult to treat with sprays. In severe cases, the bark will develop lumps that may split in frosty periods, leaving the tree open to apple canker.
CONTROL: If the aphids are noticed early, simply paint them with methylated spirits, or scrape them off individually. If larger areas are infected, spray with derris. Failing this, cut the resulting lumps out from the bark.

Below: These apples have the tell-tale scars caused by apple sawfly maggots.

Below: This apple damage is caused by a tunnelling codling moth caterpillar.

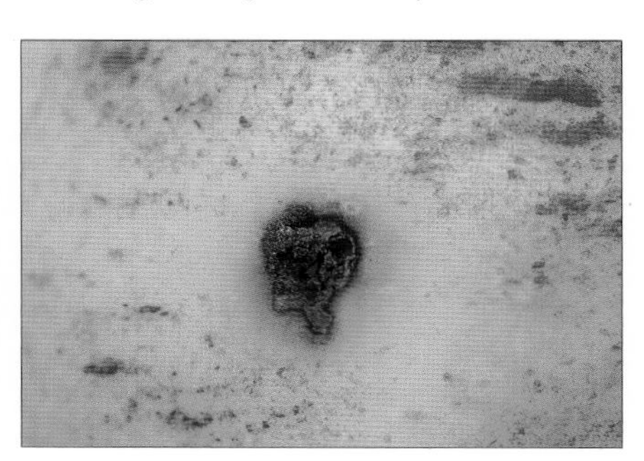

Diseases

Bacteria, viruses or fungi are the main causes of diseases in apples, often entering plants through the flowers or wounds.

Canker

A serious fungal disease (*Nectria galligena*) of apples and related fruits.
SYMPTOMS: Sunken, discoloured patches are seen on bark. The canker is first noticeable in autumn as a swelling of the bark – often at the site of a pruning wound or damaged bud. The central part of the swelling begins to die back and the bark flakes off leaving a sunken discoloured area. In summer, white fungus grows on the diseased bark, turning to a red fungus-like growth in winter.
CONTROL: Diseased patches should be cut back to healthy wood, using a knife or chisel. Burn the diseased wood. The exposed healthy wood can be painted with a canker paint. If canker is a major problem, spray with a copper-based fungicide in late summer to early autumn. Three consecutive sprayings are needed.

Mildew (powdery and downy mildew)

A fungal disease caused by a number of different pathogens.
SYMPTOMS: Light grey powdery patches appear on leaves, shoots and flowers, normally in spring. The flowers turn a creamy yellow colour and will not develop correctly. Powdery mildew remains on the surface of the plant. Downy mildew penetrates the plant, eventually killing it. Both types of mildew are prevalent in cool and over-damp conditions.
CONTROL: All infected growth should be removed and burnt – do not put it on the compost heap. If the problem persists, spray with copper fungicide. Avoid over-watering and prune to improve air circulation within the crown of the tree. Reduce the amount of fertilizer being applied.

Below: The fungal spores of apple canker surround the sunken patches on this branch.

Brown rot

A fungus (*Monilinia fructigena* and *M. laxa*) that enters fruits through wounds made by wasps, caterpillars and birds.
SYMPTOMS: Entire fruits rot and turn brown. The fruit becomes soft and grey spots of fungus grow on the browned fruit. Eventually the fruit will shrivel and fall off.
CONTROL: The disease is spread by contact, so all infected fruit, whether on the tree or on the ground, should be removed and burnt as soon as possible. If the disease has reached fruits in storage, these should also be removed and burnt and the storage area thoroughly cleaned with a disinfectant. Annual cleaning of the storage area with soda and warm water is a good preventive measure. Keep the soil and grass around the tree clear, removing leaves and other debris regularly. No chemical control is available.

Scab

A fungus (*Venturia inequalis* and *V. pirina*) that attacks apples as well as other related fruits.

Below: The light grey powdery patches on these apple leaves are caused by mildew.

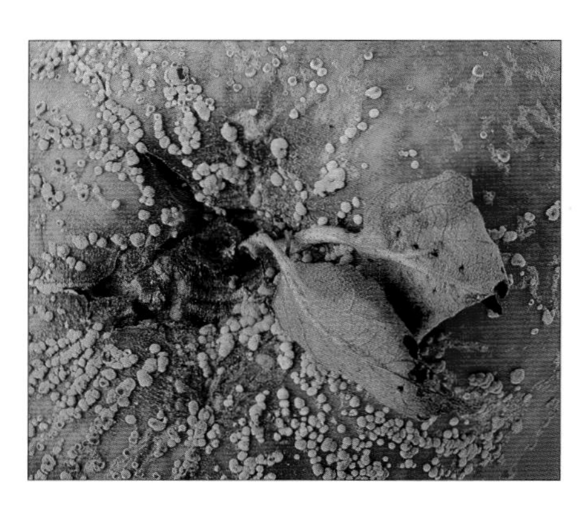

Above: This apple has turned brown and is covered with buff fungal spores of brown rot.

SYMPTOMS: Olive green or brown blotches appear on the leaves. The blotches turn browner in time and brown scabs appear on the fruit. The diseased leaves will fall early and the fruit will become increasingly covered in scabs – eventually the fruit skin will crack.
CONTROL: Remove and burn diseased fruit and leaves (including shed leaves) as soon as possible. Spores can over-winter in the fallen leaves, so clear and burn all leaves which fall in the autumn and winter. When planting new trees, avoid low-lying areas where air movement is restricted. Plant the trees in the open where they will benefit from good air circulation, avoiding damp conditions.

Granny Smith and Delicious are particularly susceptible.

Botrytis (grey mould)

A common fungal disease (*Botrytis* spp.) that can affect nearly all plants.
SYMPTOMS: Botrytis is first noticeable as brown spots, which are followed by a furry grey mould. The cause of the disease is too much dampness in cool conditions – growing plants in over-fertile conditions encourages botrytis.
CONTROL: All infected growth should be removed and burnt – do not put it on the compost heap. If the problem persists, spray with copper fungicide (Bordeaux mixture). To prevent further occurrences, improve growing conditions. Avoid over-watering and prune to improve air circulation within the crown of the tree. Reduce the amount of fertilizer being applied.

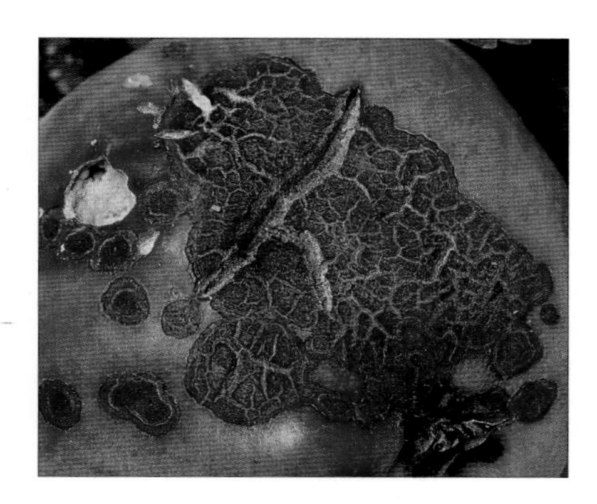

Above: Apple scab is a fungal disease that is spread by wind and rain.

Fireblight

A bacterial disease (*Erwinia amylovora*).
SYMPTOMS: Leaves turn brown and wither without being shed by the plant. Shoots can die back. Flowers wilt and young shoots wither and die. Cankers appear around the base of dead shoots in autumn. In severe cases, trees can die.
CONTROL: Cut back affected growth, cutting at least 60cm (2ft) behind the diseased material. Plant disease-resistant varieties. Spray trees when in flower with copper fungicide.

Silver leaf

A fungus (*Chondrostereum purpureum*) that enters plants through recent wounds, such a snapped branch, or often as a result of pruning during or just before a period of wet weather.
SYMPTOMS: Leaves turn silver then brown. The silver sheen is due to a poison which the fungus releases that causes the outer cells on the leaf to separate. Branches die back. Mauve fruiting bodies (brackets) appear on dead wood. The extent to which the disease has penetrated the tree can be determined by cutting off a branch (at least 2.5cm/1in in diameter) and wetting the cut surface. Affected tissue shows as a purple or brown stain.
CONTROL: Cut back all affected growth to about 15cm (6in) beyond the point where the brown or purple staining ceases. Sterilize all tools before and after use. Feed, water and mulch plants well to promote recovery. If fruiting bodies have appeared on the main trunk, dig up the tree and destroy it. To minimize risk of infection, carry out all pruning during settled, warm weather in early summer and paint all pruning wounds with a wound paint.

Physiological problems

Inappropriate growing conditions are often the source of problems. Irregular water supplies can lead to cracking of the fruits and uneven growth. Water trees well during periods of drought while the fruits are ripening. Apply a deep annual mulch around the base of trees in spring to conserve soil moisture and maintain even growth.

Bitter pit

Caused by a chemical imbalance in the tree, bitter pit indicates either a shortage of calcium or too much potassium or magnesium. This is often due to a shortage of water at a crucial time in the development of the fruits.
SYMPTOMS: Several small brown sunken pits on the surface of the apple. The flesh below the pits is also browned and tastes bitter. The number of affected areas will rise considerably during storage. Fruits showing signs of bitter pit should not be stored.
CONTROL: There is no reliable treatment once the problem has been noticed. To avoid recurrence, the following season mulch around the tree with well-rotted compost to conserve water, especially in dry periods. Do not over-fertilize.

Biennial bearing

Producing irregular crops is a common problem of apples, particularly of some varieties. It can be caused by a number of factors.
SYMPTOMS: Trees crop heavily one year then produce hardly any the following year. The habit can be set in the year after a particularly hard winter that has wiped out all the flowers so no fruit is set during the season following. The tree then bears heavily as if to compensate, thus exhausting its capacity to flower and fruit the year after that. Heavy attack by pests and diseases at flowering time can also lead to biennial bearing in the same way.
CONTROL: If a tree has cropped poorly, reduce the number of flowers in the following spring, pinching out up to nine out of ten clusters (leaving the surrounding leaves intact). This leads to a moderate crop that year, and the tree should crop normally the year after, thus breaking the pattern.

Replant disease

This disorder occurs where a plant is replaced with one of the same type that then fails to thrive.
SYMPTOMS: Plants establish poorly and roots fail to grow. The plant may die.
CONTROL: If the new plant is still alive, dig it up and transfer it to a site where apples have not already been grown. Alternatively, dig it up, then excavate a fresh hole in the same place, 60cm (2ft) across and 30cm (1ft) deep. Wash the roots under running water, then replant the tree, backfilling with fresh soil taken from another part of the garden. Use of mycorrhizal fungi when planting is believed by some gardeners to reduce the risk of replant disease.

Below: The dark blotches on these apple leaves are caused by silverleaf.

Below: The various brown sunken pits on this apple indicate a chemical imbalance.

A DIRECTORY OF APPLES

With more than 7,500 apple cultivars known to exist worldwide, it is not surprising that this well-loved fruit comes in such a diverse range of shapes and sizes.

The apples on the following directory pages are listed alphabetically, from Acme to Zoete Ermgaard, for easy reference. The selection provides an overview of as varied and as international a selection of apples as possible.

Each directory entry gives a detailed description of the shape, size, flavour, colour and texture of the fruit. The cultivars are illustrated photographically from the top, the bottom and cut in half. Some fruits are shown growing on the tree. Entries are accompanied by notes and a text panel provides information on the type, origin, parentage, flowering season and other names of each variety.

Above left: *Rossie Pippin.* ***Above right***: *Winston.*
Left: *A flourishing crop of attractive rosy-red fruit weighs down the branches of this apple tree.*

Acme

The rounded fruits, around 6cm (2¼in) across, are yellowish green, heavily flushed with crimson. The flesh is firm and creamy with a rich and fruity flavour.

Type: Dessert.
Origin: Boreham, Essex, UK, 1944.
Parentage: Worcester Pearmain x ?Rival (female) x Cox's Orange Pippin (male).
Flowering: Mid-season.

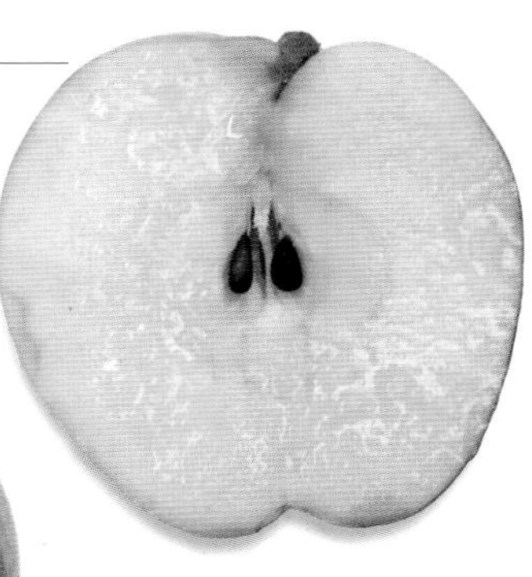

***Right**: Fruits develop red striping as they ripen.*

Note: This variety is self-sterile so a compatible pollination partner must be grown nearby to ensure fruiting.

Adam's Pearmain

This old variety produces conical fruits, around 5cm (2in) across or more, that are citrus yellow striped with red, with regular russeting. The flesh is creamy yellowish white, firm-textured, juicy and with a nutty flavour similar to Egremont Russet. This was a popular variety in the 19th century.

Type: Dessert.
Origin: Herefordshire, UK, 1826.
Parentage: Unknown.
Flowering: Mid-season.
Other names: Adamsapfel, Hanging Pearmain, Lady's Finger, Matchless, Moriker, Norfolk Pippin, Norfolk Russet, Pearmain d'Adam, Rough Pippin, Rousse du Norfolk, Russet aus Norfolk and Winter Striper Pearmain.

Note: Adam's Pearmain shows some resistance to scab. Young trees bear freely.

Alfriston

This traditional old variety has slightly knobbly, bright yellow-green fruits, often more than 7cm (2¾in) across, with a sharp, acid flavour. They are soft and coarse in texture. Fruits can be stored for several months and cook well.

Type: Culinary.
Origin: Uckfield, Sussex, UK, late 1700s; renamed Alfriston 1819.
Parentage: Unknown.
Flowering: Mid-season.
Other names: Alfreston, Freen Grove, Green Goose, Lord Gwydyr's Newton Pippin, Oldaker's New, Shepherd's Pippin and Shepherd's Seedling Pippin.

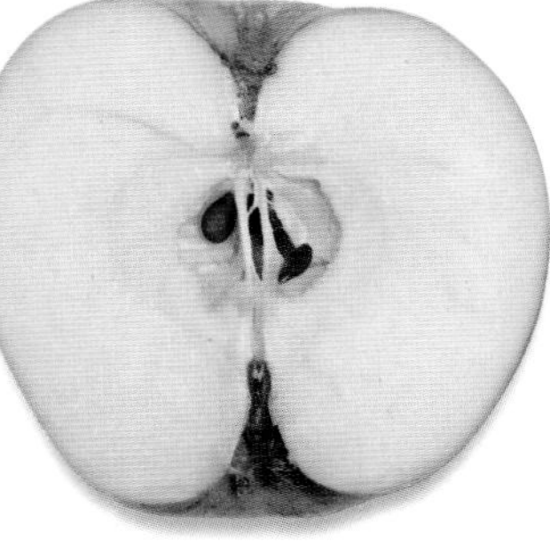

Note: Canker can be a problem on heavy soils. Trees bear heavily.

Allen's Everlasting

Type: Dessert.
Origin: Ireland, before 1864.
Parentage: Possibly a seedling of Sturmer Pippin.
Flowering: Mid-season.
Other names: Allen's Dauerapfel, Eternelle d'Allen and Harvey's Everlasting.

Slightly flattened in shape, the fruits, around 6cm (2¼in) across, are rough-skinned and greenish yellow, touched with red as they ripen and russeted over the whole surface. The flesh is firm, juicy and pale yellow with a rich flavour.

Note: Trees are not particularly vigorous. The apples store well.

Right: The skins of this variety show pronounced russeting as they ripen.

Allington Pippin

Type: Dessert.
Origin: Lincolnshire, UK, before 1844, originally as Brown's South Lincoln Beauty; renamed 1894 and introduced 1896.
Parentage: King of the Pippins (female) x Cox's Orange Pippin (male).
Flowering: Mid-season.
Other names: Allington, Allingtoner Pepping, Brown's South Lincoln Beauty, Pepin d'Allington, South Lincoln Beauty and South Lincoln Pippin.

The conical fruits, around 7cm (2¾in) across, are dull green or yellow with a red flush or red striping. The flesh is creamy white. The flavour is distinct, strong and sharp, with a bitter-sweet quality.

Note: Unripe fruits are also suitable for cooking, keeping their shape well.

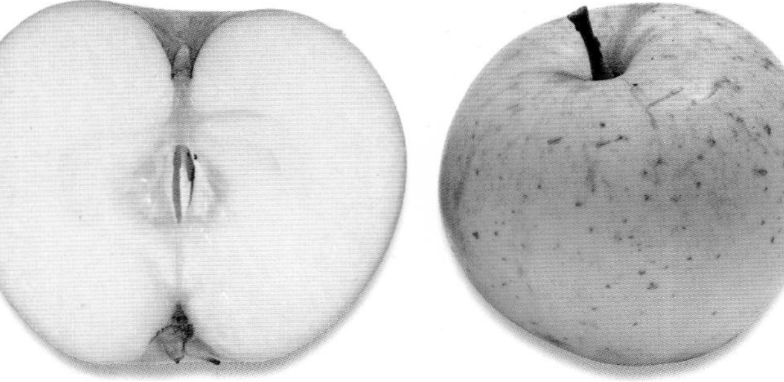

Amanishiki

Type: Dessert.
Origin: Amori Apple Experiment Station, Japan, 1936, renamed 1948.
Parentage: Ralls Janet (female) x Indo (male).
Flowering: Mid-season.

The fruits, around 5cm (2in) across, are rounded and yellowish green with a red flush. The creamy white flesh is sweet in flavour but inclined to be insipid. Fruits ripen best in areas with reliably warm summers.

Note: Its female parent is an old variety from Virginia – evidence of the importance of maintaining old varieties for breeding purposes.

Ananas Reinette

This variety produces bright golden yellow, almost cylindrical, yellow green fruits, up to 6cm (2¼in) across, with russet freckling. The flesh is white-yellow, crisp and juicy, with an intense flavour that is reminiscent of pineapples when they are ripe. The skins can turn almost orange in the sun.

Type: Dessert.
Origin: Believed to be the Netherlands, recorded 1821.
Parentage: Unknown.
Flowering: Mid-season.
Other names: Ananas, Ananasii, Ananasova reneta, Ananasrenett, Ananasz renet, d'Ananas, Reinette Ananas and Renetta Ananas.

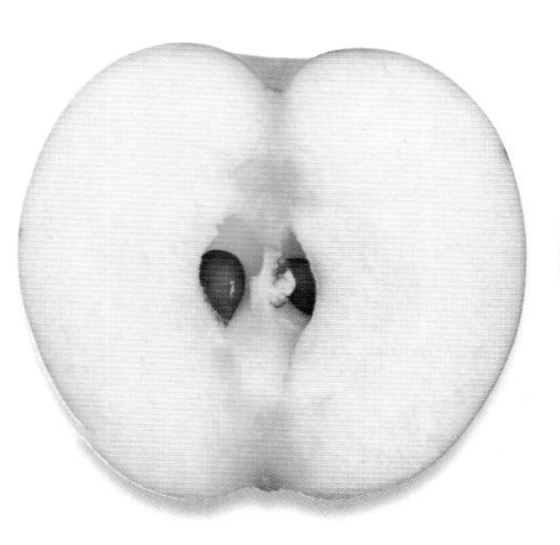

Note: Trees are suitable for training in all forms, making this a useful apple for small gardens. This is a popular apple in northern Europe.

Angyal Dezso

The fruits, of slightly irregular shape, are around 7cm (2¾in) across. They are greenish yellow with some light brown russeting. The flesh is crisp and coarse with a sweet to subacid and nutty flavour.

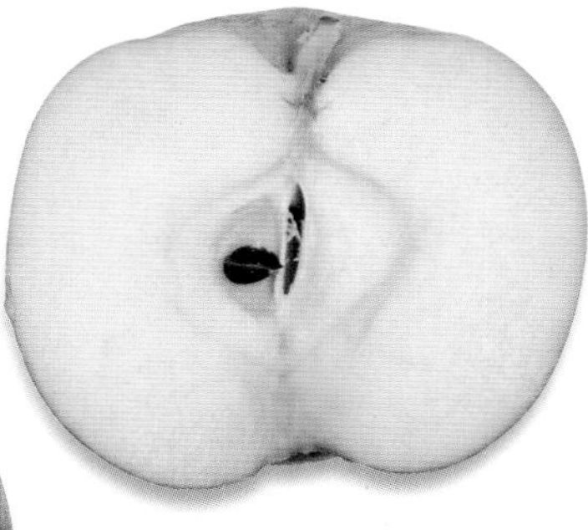

Note: Trees are moderately vigorous. They are self-sterile so need a suitable pollinator.

Type: Dessert.
Origin: Believed to be Hungary, *c.*1900.
Parentage: Unknown.
Flowering: Late.

Annie Elizabeth

This variety is unusual among cultivated apples in having maroon flowers. The large, irregular (prominently ribbed or angular), flattened fruits, to 8cm (3in) across, are bright yellow-green, flushed with red. The flesh is white. Its sweet flavour makes it one of the best apples for stewing and baking.

Type: Culinary.
Origin: Knighton, Leicester, UK, *c.*1857 (introduced *c.*1898).
Parentage: ?Blenheim Orange (female) x Unknown.
Flowering: Late.
Other names: Carter's Seedling, Sloto, Slotoaeble, Slotrable, Slotraeble, Sussex Pippin and The George.

Note: Trees show some disease resistance and are tolerant of mild, damp climates. The fruits store well.

Annurca

Type: Dessert.
Origin: Italy, 1973.
Parentage: Unknown.
Flowering: Mid-season.
Other names: Annurca Bella del Sud and Annurca Rossa del Sud.

The rounded fruits, to 5cm (2in) across, are bright yellow-green with a red flush. The creamy white flesh is sweet, crisp and juicy. This variety is of great commercial importance in Italy, and is grown almost exclusively in Campania.

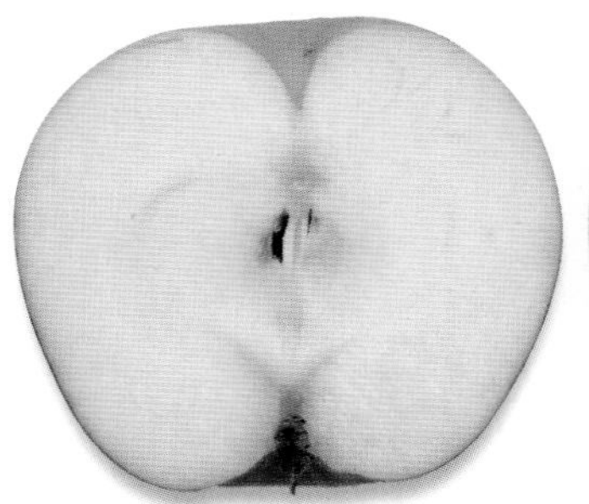

Note: In 2001, Annurca was recognized as a typical regional product by the European Union and awarded GPI (protected geographical indication) status under the name Melannurca Campana.

Right: This variety is well-known in its native Italy.

Antonovka

Type: Culinary and dessert.
Origin: A Russian variety that arose in Kursk, first recorded in 1826.
Parentage: Unknown.
Flowering: Early.
Other names: Antenovka, Antoni, Antonifka, Antonovka obyknovennaya, Antonovka prostaya, Antonovskoe yabloko, Antonowka, Antony, Bergamot, Cinnamon, Dukhovoe, German Calville, King of the Steppe, Nalivia, Possart's Nalivia, Russian Gravenstein and Vargul.

The irregular fruits, often larger than 7cm (2¾in) across, are bright yellowish green with some russeting and spotting. The creamy white flesh is crisp and juicy with an acid flavour. The fruits cook well.

Note: Trees are vigorous and hardy. The fruits cook well. This variety is sometimes used as a rootstock because of its resistance to cold.

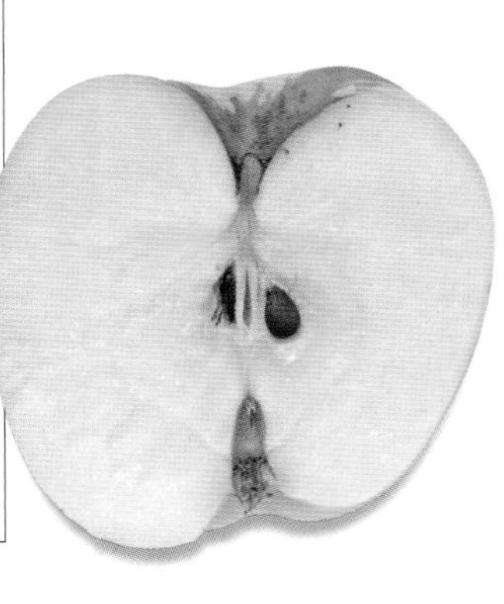

Apez Zagarra

Type: Dessert.
Origin: France, 1973.
Parentage: Unknown.
Flowering: Late.
Other name: Apez Sagarra.

The fruits, somewhat irregular in shape and around 6cm (2¼in) across, are dull green and heavily russeted. The creamy white flesh has a fairly distinct aniseed flavour.

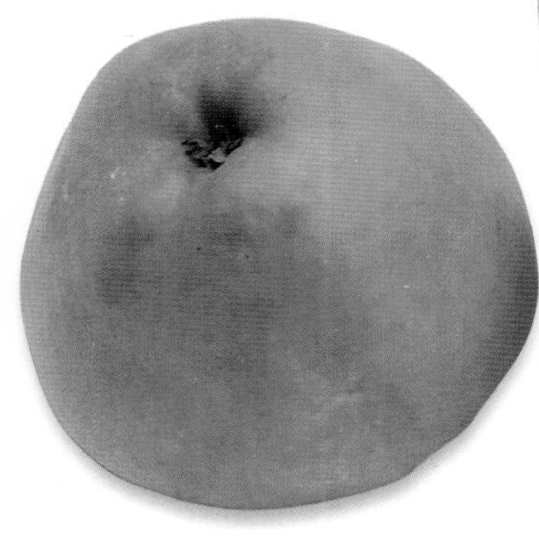

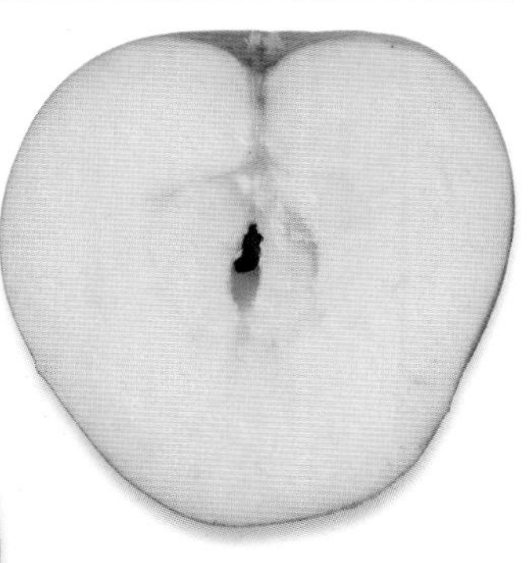

Note: This variety is local to the Basque region of France. Trees are moderately vigorous.

Ashmead's Kernel

The bright green-yellow fruits, slightly flattened and irregular in shape, are around 7cm (2¾in) across. They are flushed orange with light cinnamon brown russeting and the yellowish flesh is firm and juicy. The flavour is sweet, slightly acid and richly aromatic.

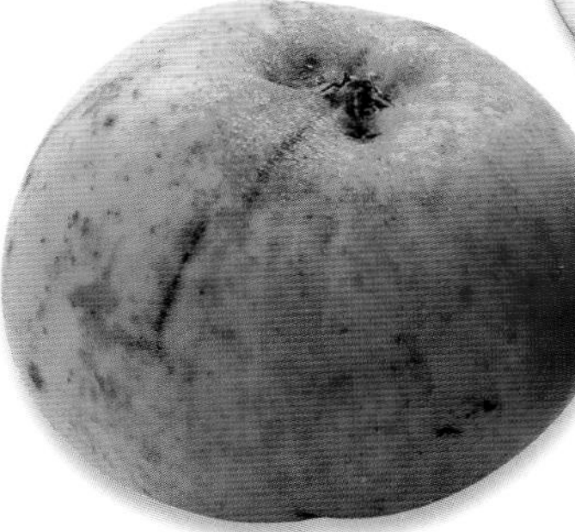

Note: This is one of the best of the old varieties. It was crossed with an unknown male to produce Improved Ashmead's Kernel, first recorded in 1883.

Type: Dessert.
Origin: Gloucester, UK, *c.*1700.
Parentage: Possibly a seedling of Nonpareil.
Flowering: Mid-season.
Other names: Aschmead's Saemling, Ashmead's Samling, Ashmead's Seedling, Dr Ashmead's Kernel, Samling von Ashmead, Semis d'Ashmead and Seyanets Ashmida.

Baldwin

This triploid variety produces uniformly large, rounded fruits, around 7cm (2¾in) across. The skins are tough, smooth, light yellow or greenish, blushed and mottled with bright red and indistinctly striped with deep carmine. Russeting is sometimes found towards the base. The yellowish flesh is coarse-textured and juicy but lacking in flavour.

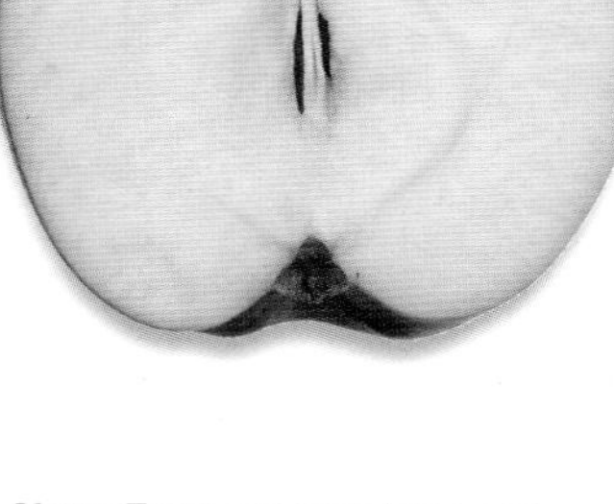

Note: Trees are very large and vigorous but biennial bearing can be a problem. This variety was formerly commercially important in the USA.

Type: Dessert.
Origin: Wilmington, MA, USA, *c.*1740, as a chance seedling (introduced *c.*1780).
Parentage: Unknown.
Flowering: Mid-season.
Other names: American Baldwin, Baldwin Rosenapfel, Baldwin's Rother Pippin, Beldvin, Butter's, Butter's Woodpecker, Butters' Red Baldwin, Calville Butter, Felch, Late Baldwin, Pecker, Pepin Rouge de Baldwin, Red Baldwin's Pippin, Steele's Red Winter and Woodpecker.

Ballarat Seedling

Note: Trees are vigorous. Fruits store well and can also be used in cooking. This variety was found in the garden of a Mrs Stewart and is sometimes informally called the Stewart apple.

The fruits are large, slightly irregular and around 8cm (3in) across. Skins are green with a red flush. The flavour is subacid and the fruits need a long hot summer to ripen fully. The flesh is coarse and hard.

Type: Dessert.
Origin: Ballarat, Victoria, Australia, early 1900s but possibly older.
Parentage: ?Dunn's Seedling (female) x Unknown.
Flowering: Mid-season.
Other names: Ballarat, Stewart's, Stewart's Ballarat Seedling and Stewart's Seedling.

Ball's Pippin

Type: Dessert.
Origin: Langley, Buckinghamshire, UK, 1923.
Parentage: Cox's Orange Pippin (female) x Sturmer Pippin (male).
Flowering: Mid-season.
Other name: Lane's Oakland Seedling.

The fruits, around 7cm (2¾in) across, are round and flattened in shape with yellowish green and russeted skin tones. The flesh is crisp and sweet in flavour.

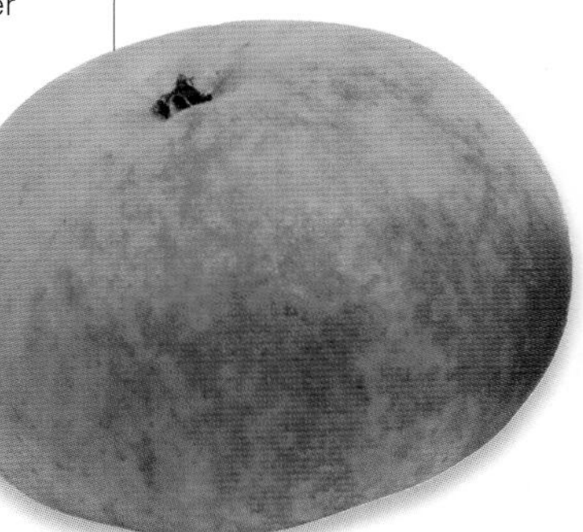

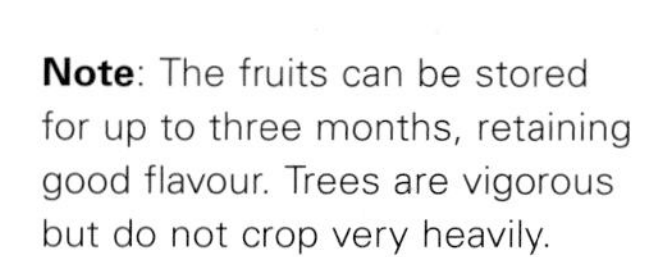

Note: The fruits can be stored for up to three months, retaining good flavour. Trees are vigorous but do not crop very heavily.

Barnack Beauty

Type: Dessert.
Origin: Barnack, Northamptonshire, UK, *c.*1840, introduced *c.*1870.
Parentage: Unknown.
Flowering: Mid-season.
Other names: Barnack, Beckford Beauty and Piekna z Barnaku.

The flushed fruits, rounded to slightly conical in shape and 5cm (2in) across or larger, are sweet, crisp and juicy with a good flavour. They are also suitable for cooking. The spring blossom is particularly attractive.

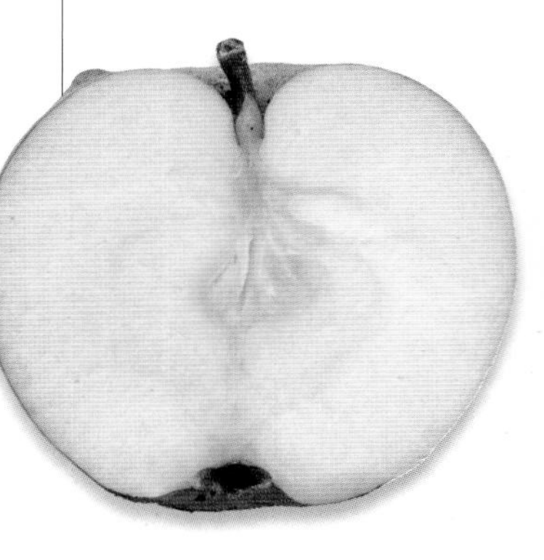

Above: The attractive Barnack Beauty has a sharp flavour and is frequently found in old orchards in the east of England.

Note: This variety may not crop well in all areas.

Barnack Orange

Type: Dessert.
Origin: Belvoir Castle, Leicestershire, UK, 1904.
Parentage: Barnack Beauty (female) x Cox's Orange Pippin (male).
Flowering: Mid-season.

The flattened fruits, to 6cm (2¼in) across, are yellow-green with a pronounced red flush. The creamy white flesh has an aromatic flavour somewhat reminiscent of Cox. The flesh is firm and rather coarse in texture.

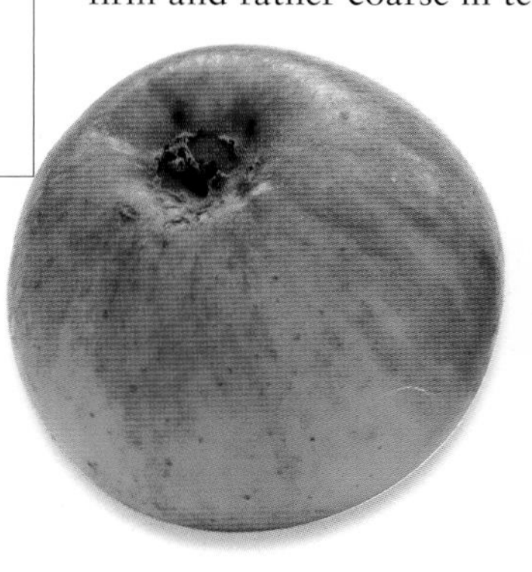

Note: Fruits can be stored for up to five months. They are similar to those of Barnack Beauty (above) but are sweeter and ripen earlier.

Bascombe Mystery

The knobbly, green fruits, around 6cm (2¼in) across, turn more yellowish as they ripen. The white flesh is firm, fine and tinged with green, with a sweet to subacid flavour.

Type: Dessert.
Origin: First recorded 1831.
Parentage: Unknown.
Flowering: Late.
Other names: Bascomb Mystery and Bascombe's Mystery.

Note: Trees are moderately vigorous with an upright to spreading habit. This variety was popular in Victorian times.

Bassard

The fruits are irregularly rounded in shape and around 6cm (2¼in) across. They are bright yellow-green with some russeting. The greenish-white flesh is firm and coarse with a rather acid flavour.

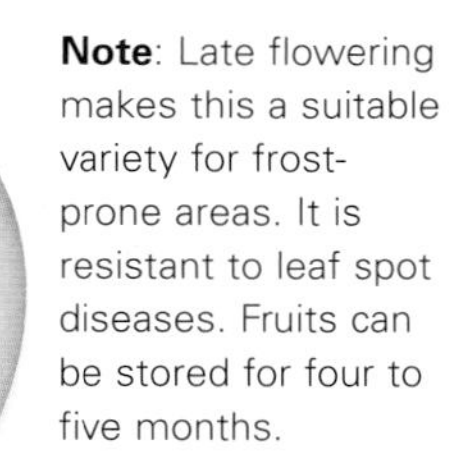

Note: Late flowering makes this a suitable variety for frost-prone areas. It is resistant to leaf spot diseases. Fruits can be stored for four to five months.

Type: Culinary.
Origin: France, first described 1948.
Parentage: Unknown.
Flowering: Late.

Baumann's Reinette

The fruits, flattened and around 7cm (2¾in) across, are greenish yellow with a brilliant red flush. The creamy white flesh is crisp and rather coarse textured and fairly juicy. The flavour is a little acid and faintly aromatic.

Type: Dessert and culinary
Origin: Belgium, 1811.
Parentage: Unknown.
Flowering: Mid-season.
Other names: Baumana, Baumanova reneta, Couronne des Dames d'Enghien, Krasnyi renet, Red Winter Reinette, Reinette de Bollwiller, Reinette rouge d'hiver de Baumann, Renet Baumann, Roter Reinette and Rothe Winter.

Note: Trees are moderately vigorous. The attractive fruits can be stored for four to five months.

Baxter's Pearmain

Type: Dessert and culinary.
Origin: Norfolk, UK, 1821.
Parentage: Unknown.
Flowering: Mid-season.

The texture of the slightly irregular fruits, around 7cm (2¾in) across, is rather coarse and dry. The skins are yellow-green with a red flush or red streaks. The creamy white flesh is slightly acid in flavour and reminiscent of Blenheim.

Note: Trees are hardy and vigorous. They bear fruit abundantly even in years when other apple varieties are cropping poorly.

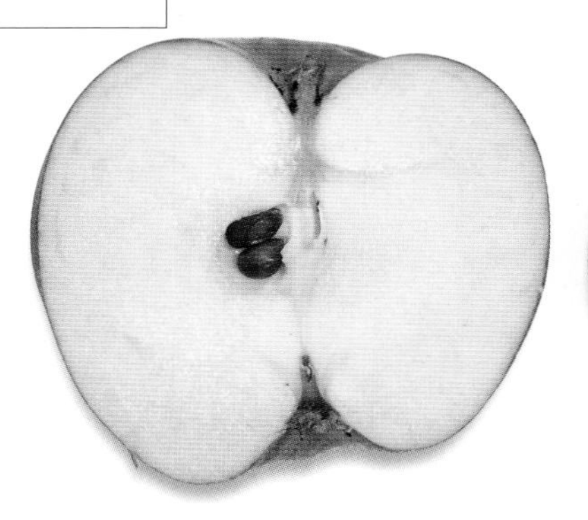

Beauty of Bath

Type: Dessert.
Origin: Bailbrook, Bath, UK, introduced *c.*1864.
Parentage: Unknown.
Flowering: Early.
Other names: Batskaya krasavitsa, Cooling's Beauty of Bath, Frumos de Bath, Krasivoe iz Bata, Schöner von Bath and Schönheit von Bath.

The slightly flattened, somewhat irregular fruits, around 6cm (2¼in) across, are bright green with a strong dark red flush and some streaking. The creamy white flesh is soft, juicy, sweet and somewhat acid with a distinctive flavour.

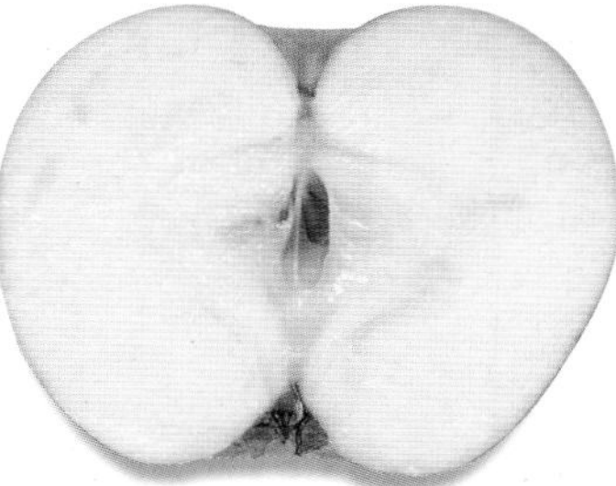

Note: This is usually one of the first apples to ripen. Trees crop heavily and show good disease resistance. Fruits do not store well and are best eaten straight from the tree. A commercially important variety in the 19th century.

Beauty of Kent

Type: Culinary.
Origin: ?England, *c.*1820.
Parentage: Unknown.
Flowering: Mid-season
Other names: Beauté de Kent, Bellezza di Kent, Countess of Warwick, Gadd's Seedling, Kentish Beauty, Kentish Broading, Kentskaya krasavitsa, Pippin Kent, Reinette Grosse d'Angleterre, Schöner aus Kent and Worling's Favourite.

The fruits, conical and irregular in shape, are around 6cm (2¼in) across and often larger. They are deep yellow, tinged with green, and show faint red patches or striping. Though edible raw, they are better cooked. The flesh, coarse in texture, is white and juicy with a sharp, subacid, though pleasant, flavour.

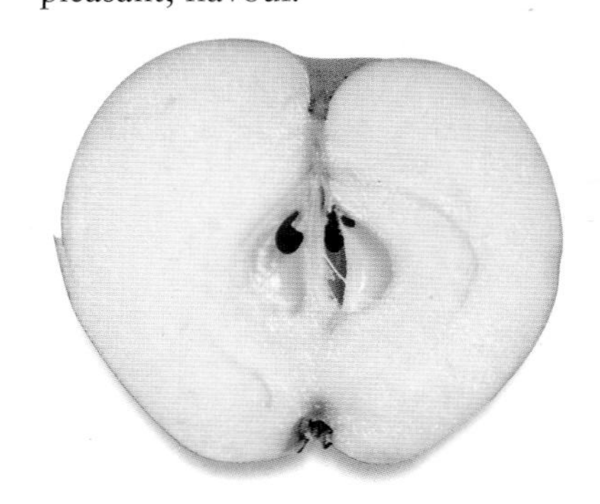

Note: This variety is suitable for growing in all areas. Canker can be a problem on heavy soils.

Beauty of Stoke

The green fruits, flushed red, are rounded and around 6cm (2¼in) across. They have a sweet flavour, unusually for a cooking apple, though the creamy white flesh is coarse and dry.

Type: Culinary.
Origin: Rufford Abbey, Nottinghamshire, UK, recorded 1889.
Parentage: Unknown.
Flowering: Mid-season.

Note: Trees are moderately vigorous. The fruits cook to a bright lemon-yellow purée.

Bedfordshire Foundling

This yellow-skinned variety has a sweet, sharp flavour. The flesh is firm, juicy and somewhat coarse-textured. The fruits, large and angular in shape, around 7cm (2¾in) across or more, cook well.

Note: Trees are moderately vigorous. The fruits can be stored for four to five months.

Type: Culinary.
Origin: ?England, *c.*1800.
Parentage: Unknown.
Flowering: Mid-season.
Other names: Bedfordshirskii Naidyonysh, Cambridge Pippin, Findling aus Bedfordshire, Mignon de Bedford, Trouvé dans le comté de Bedford and Trovatello de Bedfordshire.

Belle de Boskoop

Above: Belle de Boskoop is a versatile apple that has retained its popularity.

The fruits are lumpy, around 6cm (2¼in) across and flushed yellow-red with fawn areas. They have firm, dense, coarse-textured, white-green flesh with a pleasant, aromatic flavour and cook well. They keep well, with the flavour becoming progressively sweeter during storage.

Note: Trees are vigorous and resistant to scab. They can be slow to bear but crop well once mature. This triploid variety is very popular in continental Europe. It is vulnerable to frosts and does not do well in dry soils. Clones include Boskoop Jaune, Boskoop Rouge and Boskoop Verte.

Type: Dessert and culinary.
Origin: Boskoop, the Netherlands, *c.*1865.
Parentage: Probably a chance seedling or possibly a sport of Rechette de Montfort.
Flowering: Early.
Other names: Apfel der Zukunft, Boskoopskaya krasavitsa, Frumos de Boskoop, Gold Reinette, Piekna z Boskoop, Reinette de Montfort, Reinette Monstrueuse, Renetta di Montfort and Schoner von Boskoop.

Belle de Magny

Type: Dessert.
Origin: France, recorded 1888.
Parentage: Unknown.
Flowering: Late.
Other names: Bell de Mani, Belle de Magni and Krasavitsa iz Mani.

The fruits, irregular in shape and around 5cm (2in) across, are pale yellow-green with a pinkish red flush. They have fine, softish, creamy white flesh with a rich and subacid flavour.

Note: This variety, grown in Louis XIV's orchards, is self-fertile. It is susceptible to aphids and scab.

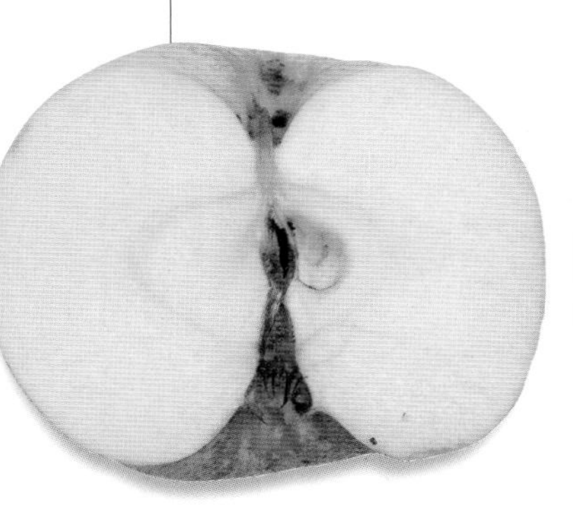

Belle de Tours

Type: Culinary.
Origin: France (Indre and Loire), 1947.
Parentage: Unknown.
Flowering: Late.
Other name: Lambron.

Fruits have crisp, juicy, white flesh with an acid flavour. Skins are bright green to whitish yellow, sometimes with a reddish blush. They are roughly conical and irregular.

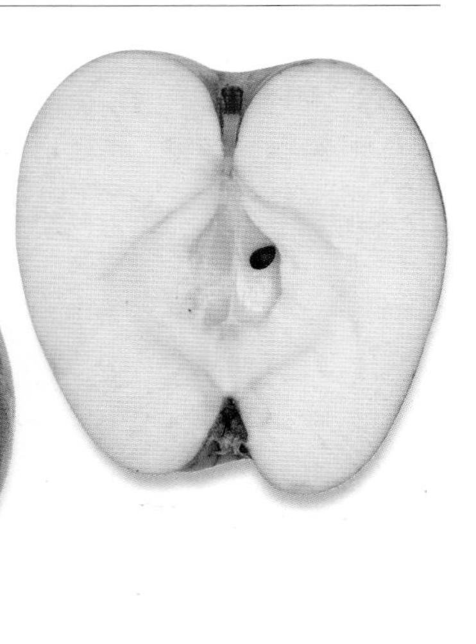

Note: This variety is increasingly rare in culivation, though old specimens are still being found in its place of origin. It is usually grown as a free-standing tree rather than being trained into a particular form.

Belledge Pippin

Type: Dessert.
Origin: Derbyshire, UK, first recorded 1818.
Parentage: Unknown.
Flowering: Mid-season.

The small, rounded, regularly shaped fruits, to 5cm (2in) across, are bright yellow-green in colour (turning yellow as they ripen) with some darker spotting. Russeting can appear as grey dots. They have coarse, soft, whitish to greenish-yellow flesh with an acid flavour.

Note: Fruits are also suitable for cooking.

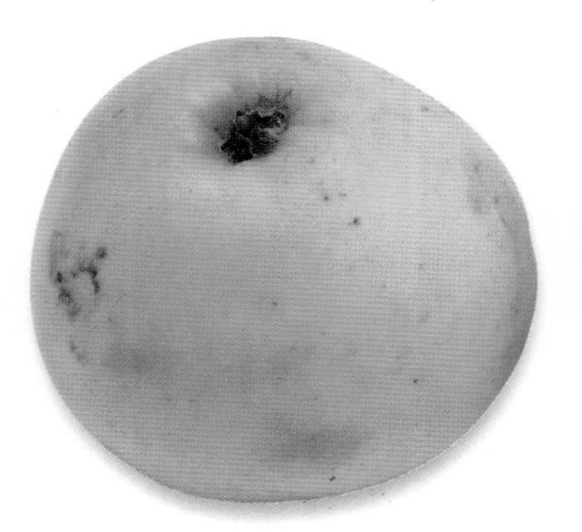

Ben's Red

The flattened fruits, around 6cm (2¼in) across, are bright yellow-green with a strong pinkish to dark-red flush and some flecking and striping. The creamy white flesh is crisp, dry and coarse with a sweet strawberry-like flavour.

Type: Dessert.
Origin: Trannack, Cornwall, UK, *c.*1830.
Parentage: ?Devonshire Quarrenden (female) x Farleigh Pippin (male).
Flowering: Early.

Note: The fruits are good for juicing.

Beregi Sóvari

The somewhat flattened fruits, around 5cm (2in) across, are bright yellowish green with some spotting and russeting. The creamy white flesh is firm and fine with a sweet flavour. This variety is believed to be a sport of Nemes Sovari Alma.

Type: Dessert.
Origin: Hungary, recorded 1900.
Parentage: Unknown.
Flowering: Mid-season.

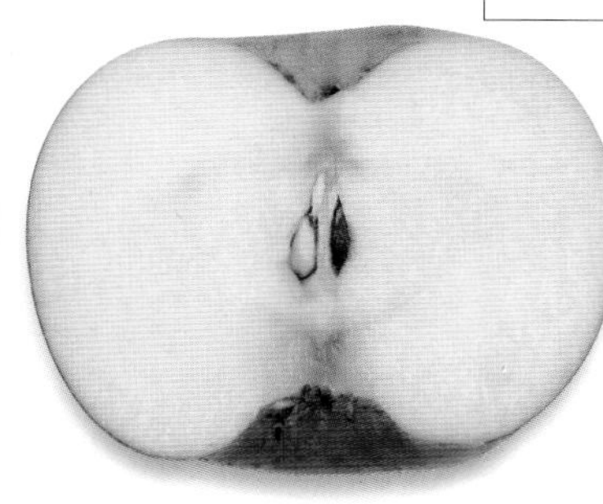

Note: Trials in Hungary have indicated susceptibility to fireblight, making this an unsuitable variety for breeding programmes.

Bess Pool

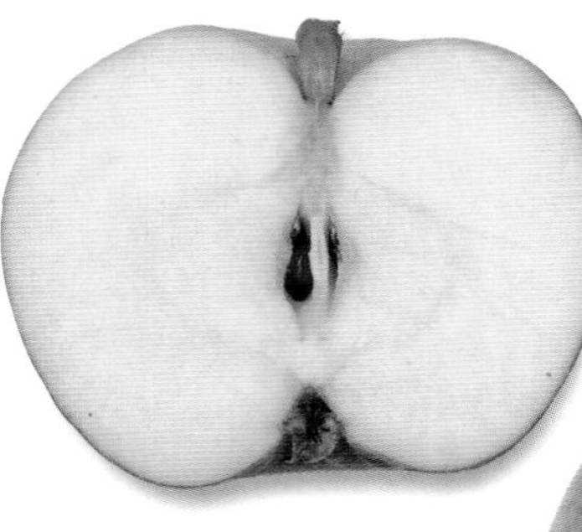

The rounded, slightly irregular fruits, around 6cm (2¼in) across or more, are yellowish green with a red, sometimes striped, flush. They have a rather dry, slightly coarse-textured, white flesh with a sweet and pleasant flavour. There is sometimes a red staining just under the skin.

Type: Dessert.
Origin: Nottinghamshire, UK, first recorded 1824.
Parentage: Unknown.
Flowering: Late.
Other names: Black Blenheim, Muskierte gelbe Reinette, Red Rice, Ronald's Besspool, Stradbroke Pippin and Walsgrove Blenheim.

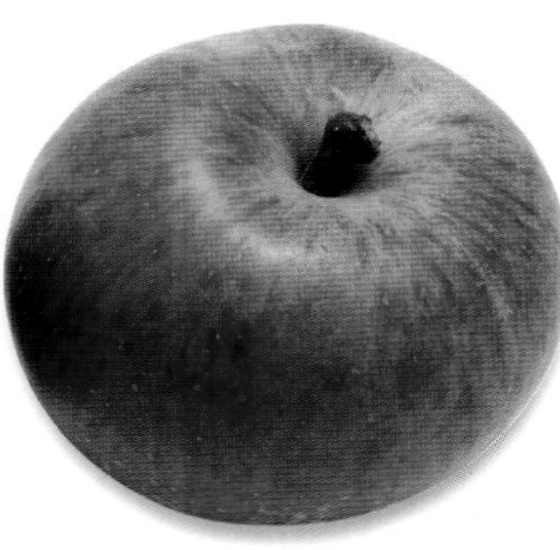

Note: Late flowering makes this variety useful in areas prone to spring frosts.

Bismarck

Type: Culinary.
Origin: Variously reported as Bismarck, Tasmania; Carisbrooke, Australia; and Canterbury, New Zealand; before 1887.
Parentage: Unknown.
Flowering: Mid-season.
Other names: Bismarckapfel, Bismarckapfel aus Neuseeland, Bismarckovo, Fürst Bismarck, Pomme Bismarck and Prince Bismarck.

The fruits are around 6cm (2¼in) across and often larger on young or regularly pruned trees. Skins are yellow-green with a strong red flush and some striping. Fruits have firm, fine-textured, juicy flesh with an acid flavour.

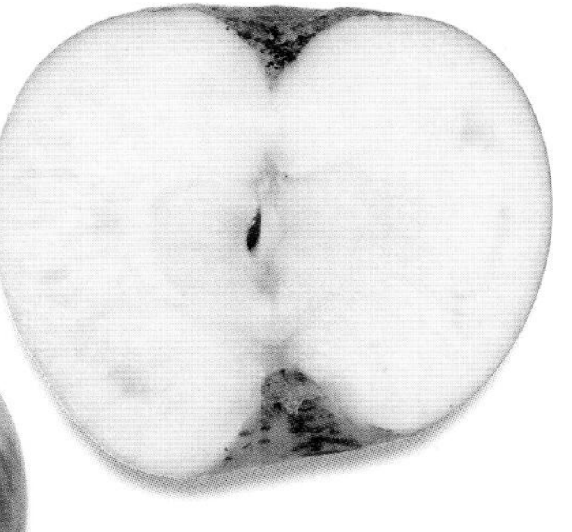

Note: This variety, named after Prince Bismarck, a German politician, is a useful general-purpose cooker. The flesh collapses on cooking. Trees are large and crop well.

Blaxtayman

Type: Dessert.
Origin: Wenatchee, Washington, USA, 1926 (introduced 1930).
Parentage: Winesap (female) x Unknown.
Flowering: Mid-season.
Other names: Nured Stayman and Stayman's Winesap.

The attractive fruits, around 6cm (2¼in) across, are yellow-green with a generally even, solid red flush. They have a very juicy, softish, yellowish flesh with a sweet flavour.

Note: This variety is a sport of Stayman Winesap, the skins of the fruits having a more even and solid flush.

Blenheim Orange

Type: Dessert and culinary.
Origin: Woodstock, Oxfordshire, UK, *c.*1740.
Parentage: Unknown.
Flowering: Mid-season.
Other names: Beauty of Dumbleton, Belle d'Angers, Blenhaimska zlatna reneta, Königin Victoria, Lucius Apfel, Northampton, Prince de Galles, Prince of Wales, Reinette dorée de Blenheim, Renet Zolotoi Blengeimskii, Ward's Pippin, Woodstock and Zlota reneta Blenheimska.

This classic variety has attractive, flattened, yellow fruits, around 7cm (2¾in) across, flushed red with fine russeting. Carried in abundance, they have a very distinctive, rich, nutty flavour. They have yellowish white, somewhat fine-textured and rather dry flesh which rapidly becomes mealy.

Note: Fruits cook well to a stiff purée. Trees, which are very vigorous, are prone to biennial bearing.

Blue Pearmain

The slightly conical, red fruits, around 6cm (2¼in) across, have a distinct bluish bloom and/or purplish red striping – hence the name. The flavour is mild, sweet, rich and aromatic. The texture is soft, rather dry and coarse.

Note: Fruits can shrivel in storage but retain good flavour. This variety was widely grown in New England in the 19th century.

Type: Dessert.
Origin: ?USA, early 1800s.
Parentage: Unknown.
Flowering: Mid-season.

Bohnapfel

The rounded fruits, around 5cm (2in) across, are light green with an orange-brown flush and striping. The skin can be tough. They have firm, coarse, yellowish white flesh with a subacid to slightly sweet flavour.

Note: This variety, a triploid, is suitable for juicing. Trees grow well at high altitudes and show good resistance to wind and frosts. There are numerous clones.

Type: Dessert and culinary.
Origin: Rhineland, Germany, late 1700s.
Parentage: Unknown.
Flowering: Mid-season.
Other names: Anhalter, Bobovoe bolsoe, Ferro rosso, Grochowka, Gros Bohnapfel, Grosser Rheinischer Bohnapfel, Jackerle, Nagy Bohn alma, Pomme Bohn, Pomme Haricot, Reinskoe bobovoe, Salzhauser Rheinischer, Strymka, Wax Apple and Weisser Bohnapfel.

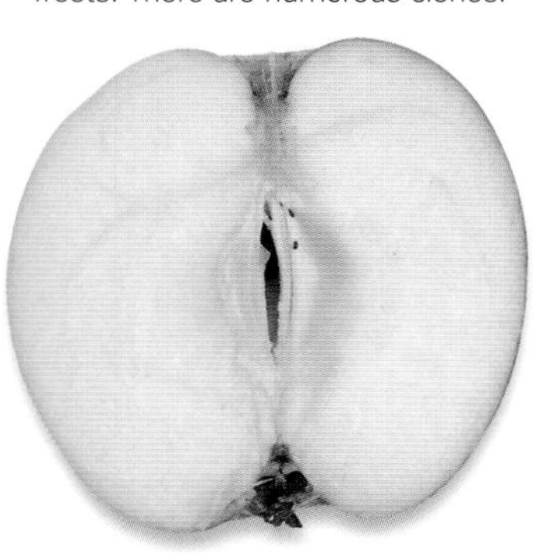

Boïken

The fruits, which are around 6cm (2¼in) across, have firm, fine-textured, juicy, acid flesh but with very little flavour. The smooth skins are yellowish, marked with pinkish red. The variety is named after Dikewarden Boïke.

Note: Trees are vigorous and show good resistance to frost but biennial bearing may be a problem. Young trees fruit well. The fruits store successfully.

Type: Dessert.
Origin: Bremen, Germany, known since 1828.
Parentage: Unknown.
Flowering: Mid-season.
Other names: Beuken, Birkin, Boiken Apfel, Bolken, Jablko Boikovo, Pomme Boiken and Zlotka Boikena.

Bolero

Type: Dessert.
Origin: Kent, UK, 1976.
Parentage: Developed from Wijcik, a sport of McIntosh.
Flowering: Mid-season.
Other name: Tuscan.

The rounded fruits, around 6cm (2¼in) across, are green with a yellow blush. The flesh is crisp and juicy with a sweet flavour. The flesh is similar to that of Granny Smith.

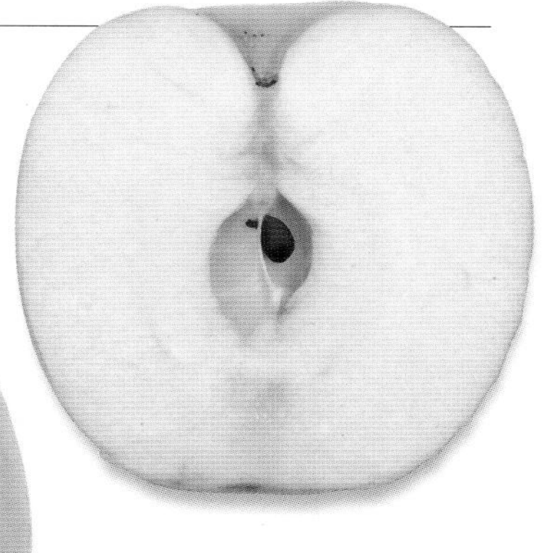

Note: This variety naturally grows as a column rather than spreading, so is useful in confined spaces. Trees show some signs of self-fertility.

Bonne Hotture

Type: Dessert.
Origin: ?Maine-et-Loire, France, recorded 1867.
Parentage: Unknown.
Flowering: Mid-season
Other names: Bonne Auture, Bonne Hoture, Bonne-Auture and de Bonne-Hotture.

The slightly flattened fruits, to 5cm (2in) across, are bright green with amber markings and conspicuous spotting. The flesh is crisp, creamy green and sweet and nutty in flavour, with a texture similar to that of a pear.

Note: Trees develop a broad crown with dense foliage. Fruits can be stored for five months or more.

Bonnet Carré

Type: Dessert.
Origin: France, recorded 1948, though believed to be much older.
Parentage: Unknown.
Flowering: Late.
Other names: Admirable blanche, Belle Dunoise, Blanche de Zürich, Calville Blanche, Calvine, Cotogna, Fraise d'Hiver, Framboise d'Hiver, Glace, Melonne, Niger, Reinette Côtelée, Taponelle and Weiss Zürich.

The fruits, angular and somewhat irregular in shape, are around 6cm (2¼in) across. They have soft, yellowish white flesh, with a sweet, perfumed flavour.

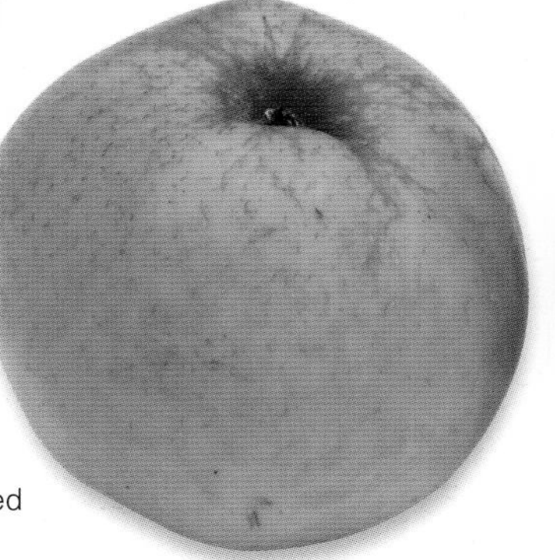

Note: This variety is best trained as an espalier. Fruits keep well.

Braddick Nonpareil

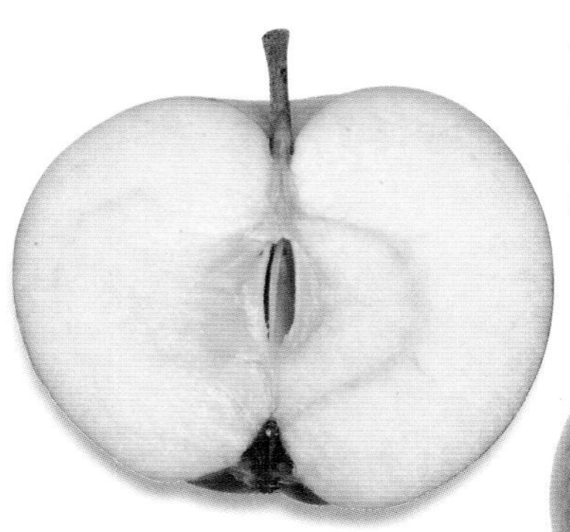

The rounded fruits, somewhat flattened, are around 6cm (2¼in) across. The skins are bright green with a strong red flush and some russeting. The firm, creamy white flesh is sugary and has a sweet-sharp flavour.

Type: Dessert.
Origin: Thames Ditton, Surrey, UK, before 1818.
Parentage: Unknown.
Flowering: Mid-season.
Other names: Braddicks Sondergleichen, Ditton Nonpareil, Ditton Pippin, Lincolnshire Reinette and Nonpareille de Braddick.

Note: The fruits store well, the flavour becoming somewhat sweeter. Trees, which crop freely, are suitable for esaplier training.

Bramley's Seedling

More commonly referred to as just Bramley, this classic English triploid variety has waxy-skinned, sometimes flattened, irregular, green fruits, up to 8cm (3in) across or more. They have an acid flavour. The flesh collapses on cooking.

Type: Culinary.
Origin: Southwell, Nottinghamshire, UK, between 1809 and 1813 (introduced 1865).
Parentage: Unknown.
Flowering: Mid-season.
Other names: Bramley and Bramleys Samling.

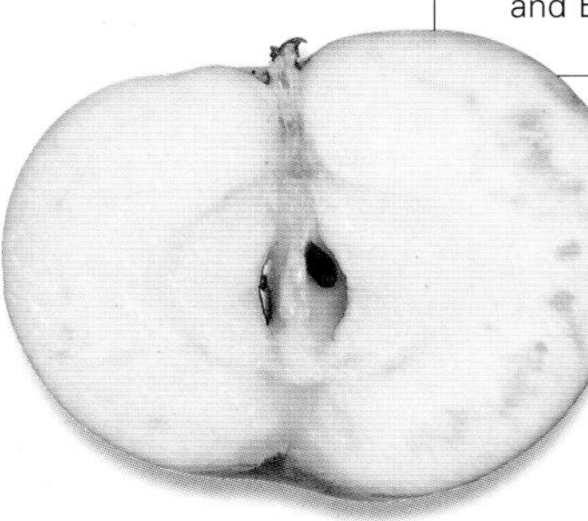

Note: Resistant to scab and mildew, this is the most popular cooking apple in the UK. Trees bear heavily.

Brenchley Pippin

This variety produces rounded, greenish yellow-skinned fruits that are around 6cm (2¼in) across. They can show areas of brownish orange strewed with russet dots. They have a sweet, aromatic, fruity flavour. The yellowish flesh, tinged green, is tender in texture, juicy and sweet.

Type: Dessert.
Origin: Brenchley, Kent, UK, 1884.
Parentage: Unknown.
Flowering: Mid-season.

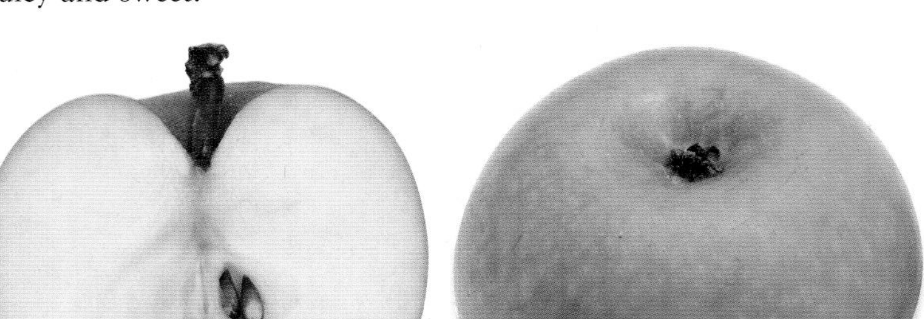

Note: The fruits store well over a long period. This apple has been widely cultivated in Kent, England.

Brownlees' Russet

Type: Dessert
Origin: Hemel Hempstead, Hertfordshire, UK, *c.*1848.
Parentage: Unknown.
Flowering: Mid-season.
Other names: Brownlee Russet, Brownlees, Brownlees graue Reinette, Brownlees-Reinette, Brownley's Russet, Reinette grise de Brownlees and Renet seryi Braunlis.

The slightly irregular fruits of this apple, around 6cm (2¼in) across, are heavily russeted with brown-green. They are juicy and have a somewhat acid, sweet-sharp but pleasantly nutty flavour. The flesh is fine-grained, crisp and greenish white.

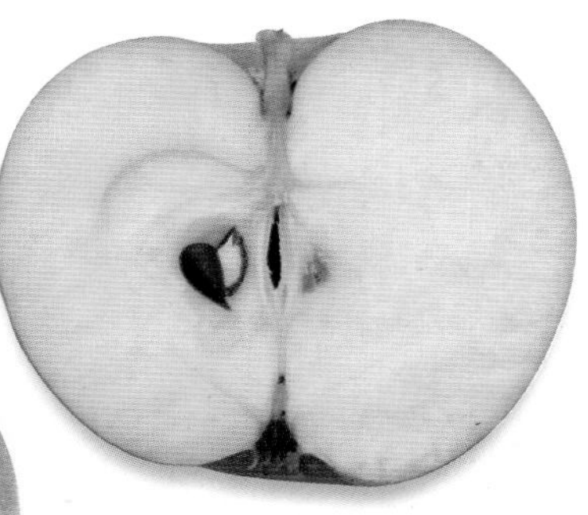

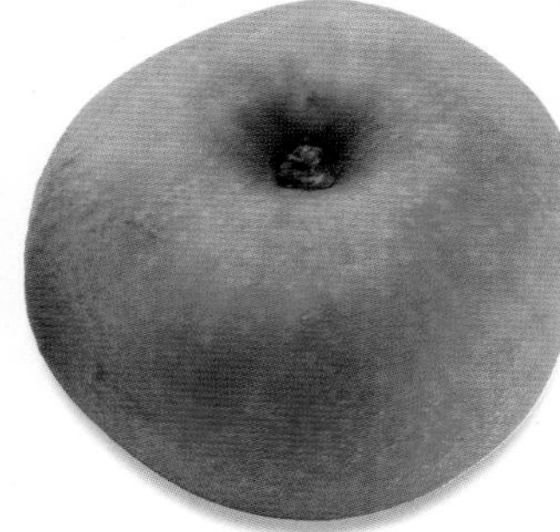

Note: This variety is also suitable for cooking. Trees are compact with attractive deep pink flowers. Fruits store well but tend to shrivel.

Burr Knot

Type: Culinary.
Origin: UK, first recorded 1818.
Parentage: Unknown.
Flowering: Mid-season.
Other name: Bute's Walking-stick.

The skins of this variety are lemon yellow with a red blush and a few russet dots. The fruits, somewhat irregular in shape and around 6cm (2¼in) across, are of no particular merit. They have soft, coarse-textured, rather dry flesh with an acid flavour.

Note: Trees are grown for their botanical and historic interests: the stems produce numerous burrs and cuttings taken from these root freely.

Cagarlaou

Type: Dessert.
Origin: Lozère, France, 1947.
Parentage: Unknown.
Flowering: Late.

The slightly conical fruits, around 6cm (2¼in) across, are bright yellowish green with a strong red flush. The creamy white flesh has a sweet, perfumed flavour.

Note: This variety, which is very hardy, is normally grown as a free-standing tree rather than being trained. The fruits store well.

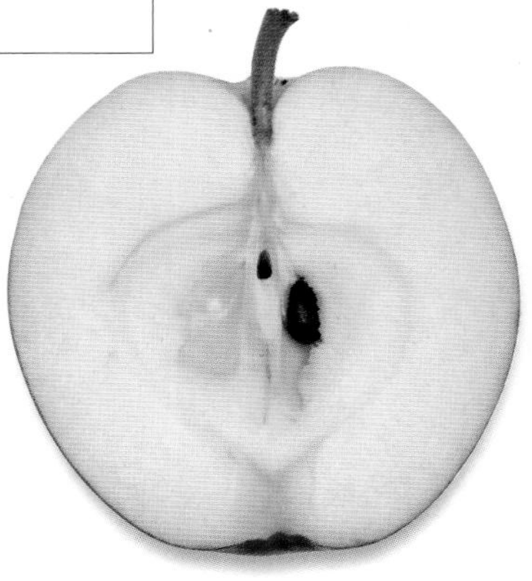

Calville Blanc d'Hiver

The fruits are knobbly and of irregular shape and around 6cm (2¼in) across. They are dull yellowish green, becoming brighter as they ripen, with some darker spotting. The creamy white flesh is soft, fine-textured and juicy with a sweet, aromatic flavour.

Note: Fruits may fail to ripen fully in cold areas unless trees are trained against a warm wall. They store well over winter and keep their shape when cooked.

Type: Dessert and culinary.
Origin: Europe, probably France or Germany, recorded 1598.
Parentage: Unknown.
Flowering: Mid-season.
Other names: a Frire, Admirable Blanche, Blanche de Zurich, Calville Acoute, Calville de Gascogne, Calville de Paris, Cotogna, Eggerling, Framboise d'Hiver, Melonne, Niger, Ostenapfel, Paris Apple, Pomme de Fraise, Pomme de Glace, Reinette a Cotes, Sternreinette, Tapounelle, Vit Vinterkalvill and Winter White Calville.

Calville de Maussion

The knobbly, uneven fruits are yellowish green with darker spotting and around 7cm (2¾in) across. The whitish flesh is sweet, subacid and aromatic in flavour.

Note: Trees are extremely vigorous. This variety is an excellent winter apple.

Type: Dessert.
Origin: France, recorded 1870.
Parentage: Unknown.
Flowering: Early.
Other names: Calleville de Maussion, Calville Maussion, Kalvil Mossion and Maussion's Calville.

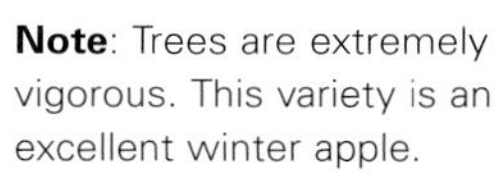

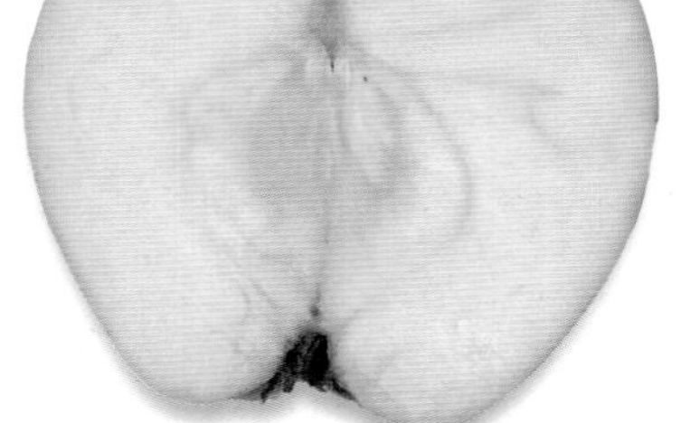

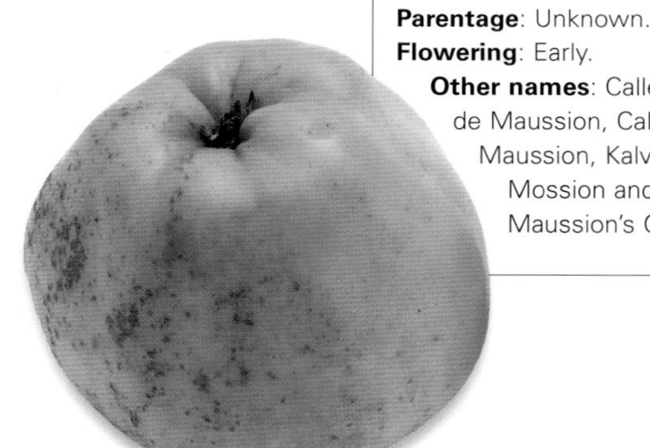

Calville Rouge D'Hiver

The rounded to slightly conical, somewhat thick-skinned, irregular fruits, around 6cm (2¼in) across, are bright yellow-green with a strong bright red flush and yellow dots. The creamy white flesh is rather soft and moderately juicy with a somewhat sweet flavour.

Note: This variety is vigorous and hardy, performing well in cold situations. It does best in fertile soil. The fruits can sometimes be small.

Right: As its French name suggests, Calville Rouge d'Hiver is a good winter apple.

Type: Culinary.
Origin: Possibly Brittany, France, recorded *c.*1600.
Parentage: Unknown.
Flowering: Mid-season.
Other names: Achte Rote Winter Calville, Blutroter Calville, Calville d'Anjou, Calville Imperiale, Coeur de Boeuf, Cushman's Black, Gallwill Rusch, Le Général, Passe Pomme d'Hiver, Rambour Turc, Red Winter Calville, Roode Paasch, Roter Winter Calville and Winter Red Calville.

Captain Kidd

Type: Dessert.
Origin: Twyford, Hawkes Bay, NZ, 1962 (introduced 1969).
Parentage: Cox's Orange Pippin (female) x Delicious (male).
Flowering: Mid-season.

The fruits are slightly conical and around 6cm (2¼in) across. They are heavily flushed red and can show some russeting. The white flesh is crisp, sweet and juicy with a rich, aromatic flavour.

Note: This variety is a more highly coloured sport of Kidd's Orange Red.

Right: *Captain Kidd can be relied on to fruit well.*

Carrara Brusca

Type: Dessert.
Origin: Italy, 1958.
Parentage: Unknown.
Flowering: Mid-season.

The flattened fruits can be up to 7cm (2¾in) across or more. They are bright green with a red flush and some russeting. The yellowish flesh is firm and crisp with a subacid flavour.

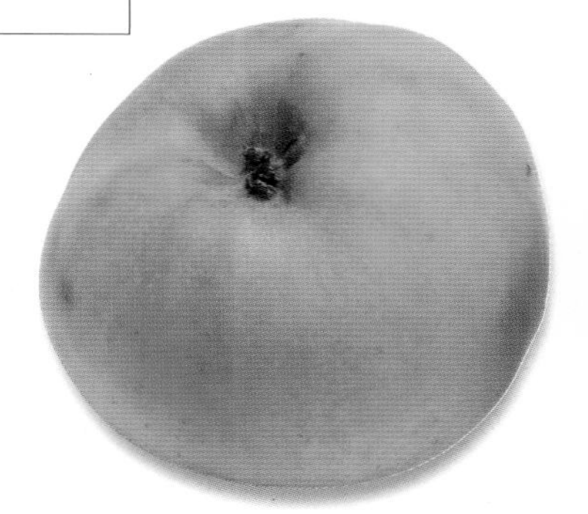

Note: This variety is rare in cultivation. Cankers have been reported.

Catshead

Type: Culinary.
Origin: England, UK, before 1600s.
Parentage: Unknown.
Flowering: Mid-season.
Other names: Apfelmuser, Coustard, Crede's Grosser, Wilhelm's Apfel, Duke of York, Green Codlin, Green Costard, Grenadier, Herefordshire Goose, Herrenapfel, Katzenkopf, Loggerhead, Monstrous Pigs Snout, Pomme de Royal Costard, Schafsnase, Stoke Leadington, Tankard, Terwin's Goliat, Tête de Seigneur and Tête du Chat.

The fruits of this unusual variety have a very distinctive, angular shape and are up to 7cm (2¾in) across or more. Fruits have a green skin. The texture of the whitish flesh is coarse and rather dry and the taste is sharp and subacid.

Note: The fruits cook to a firm, sharp-flavoured purée. Young trees may not crop well.

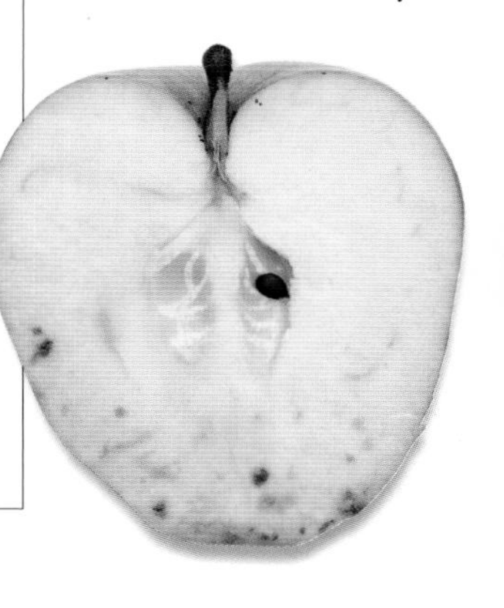

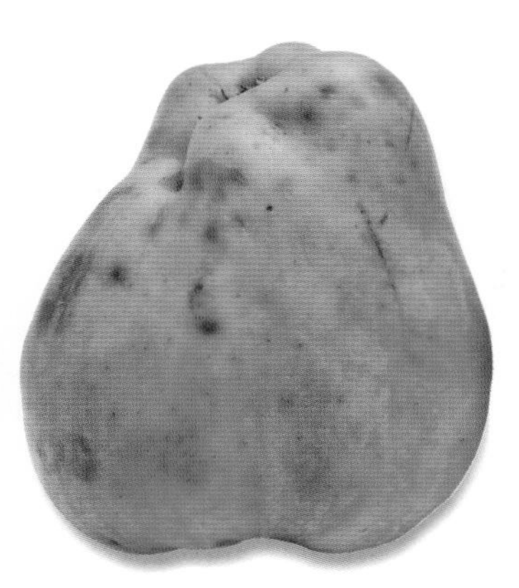

Cavallotta

The rounded but slightly irregular fruits are around 6cm (2¼in) across. The skins are bright yellowish green and the flesh is creamy white and firm with an acid flavour.

Type: Dessert.
Origin: Italy, 1958.
Parentage: Unknown.
Flowering: Mid-season.

Note: This variety is spur-fruiting.

Cheddar Cross

The conical fruits are around 7cm (2¾in) across and pale dull yellowish green with a red flush. The firm, fine-textured, white flesh is somewhat acid with little flavour.

Note: This variety is resistant to scab. Growth is dense and trees readily form spurs.

Type: Dessert.
Origin: Long Ashton Research Station, Bristol, UK, 1916 (introduced 1949).
Parentage: Allington Pippin (female) x Star of Devon (male).
Flowering: Early.

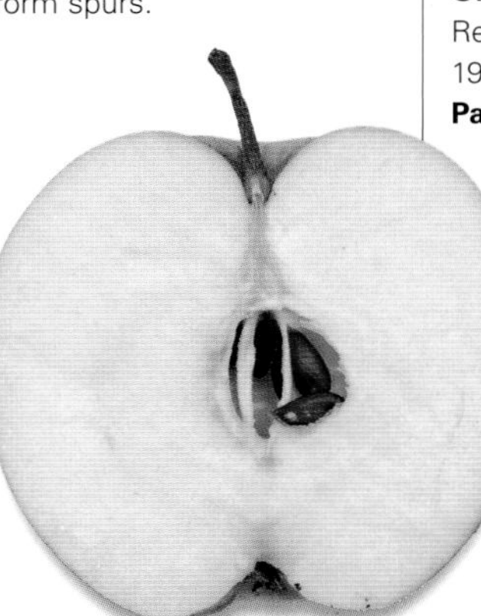

Chelmsford Wonder

The flattened, somewhat irregular fruits are around 7cm (2¾in) across. They are dull greenish yellow with red striping. The near white flesh is firm, juicy and fine-textured with a subacid flavour.

Type: Culinary.
Origin: Chelmsford, Essex, UK, *c.*1870 (introduced 1890).
Parentage: Unknown.
Flowering: Mid-season.
Other names: Chudo Shelmsforda, Merveille de Chelmsford and Wunder von Chelmsford.

Note: Fruits store well for several months. The flavour remains acidic after cooking.

Chivers Delight

Type: Dessert.
Origin: Histon, Cambridgeshire, UK, 1936.
Parentage: Unknown (possibly involving Cox's Orange Pippin).
Flowering: Mid-season.

The rounded fruits are around 6cm (2¼in) across. They have bright green skins with a strong red flush. The creamy white flesh is crisp and juicy with a sweet, pleasant flavour.

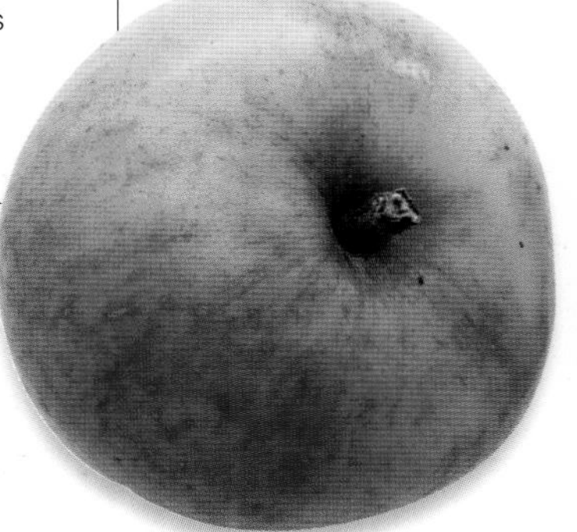

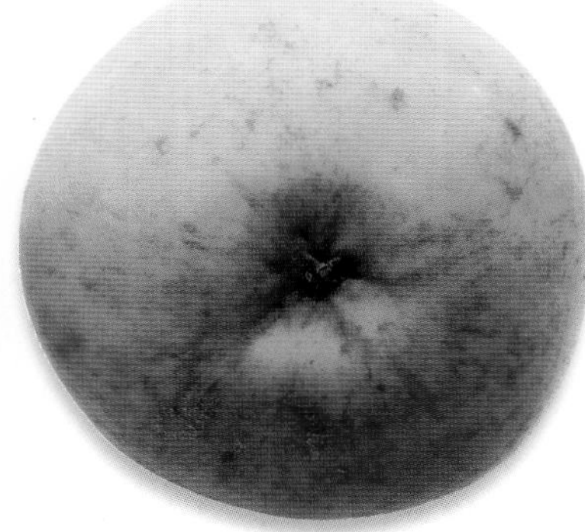

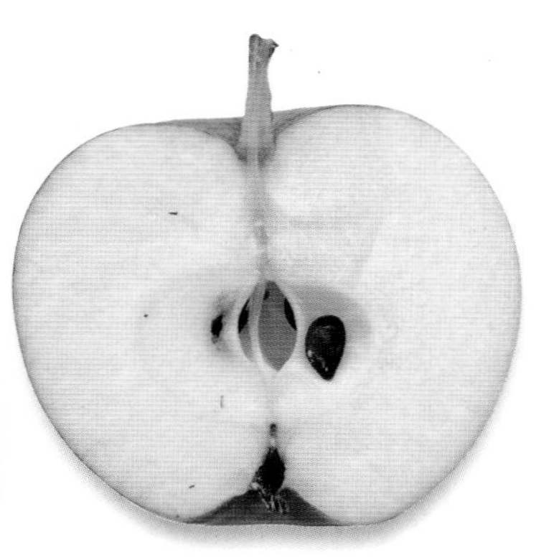

Note: Susceptible to canker. Fruits store well. This variety was popular during the early part of the 20th century.

Christmas Pearmain

Type: Dessert.
Origin: Kent, UK, first recorded 1893.
Parentage: Unknown.
Flowering: Early.
Other name: Bunyard's Christmas Pearmain.

The rounded fruits, around 6cm (2¼in) across, are green with a dark red flush. The flesh is yellowish white, crisp and juicy and has a pleasant, sweet flavour.

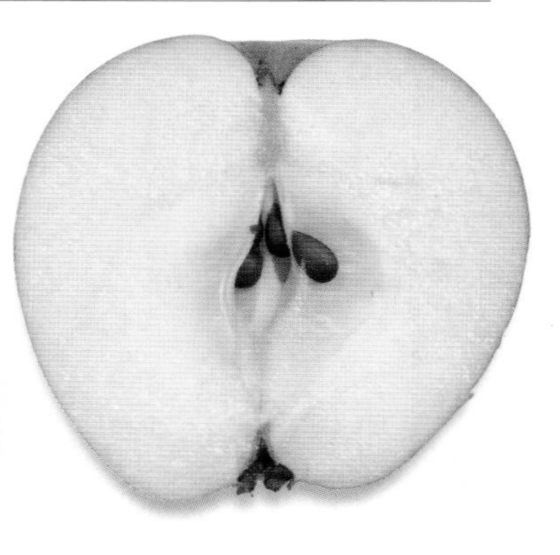

Note: Trees are generally healthy and crop well.

Claygate Pearmain

Type: Dessert.
Origin: Claygate, Surrey, UK, exhibited 1821.
Parentage: Unknown.
Flowering: Mid-season.
Other names: Archerfield Pearmain, Bradley's Pearmain, Brown's Pippin, Doncaster Pearmain, Empress Eugenie, Formosa Nonpareil, Formosa Pippin, Fowler's Pippin, Mason's Ribston Pearmain, Parmen Kleigatskii, Pearmain de Claygate, Pomme de Claygate, Ribston Pearmain, Summer Pearmain and Winter Pearmain.

The rounded fruits, often more than 6cm (2¼in) across, are dull yellow-green with a pinkish red flush and striping and some thin russeting and dotting. The yellowish white flesh is firm, crisp, rather coarse-textured and juicy with a rich aromatic flavour.

Note: Fruits can be stored for around four months. Trees are not vigorous, staying neat and compact, but crop freely, showing some resistance to scab. This variety is a good choice for a small garden.

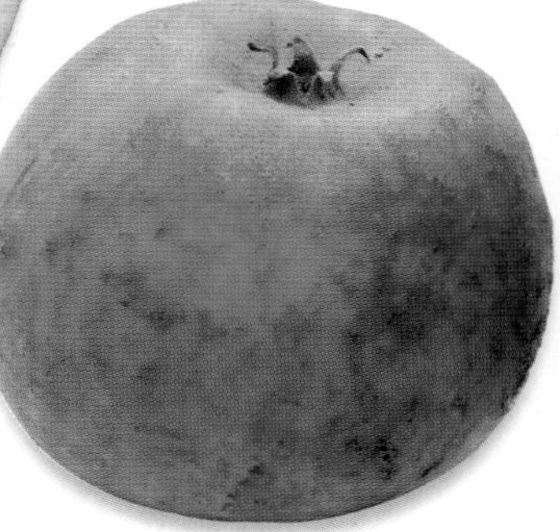

Clemens

The slightly flattened, irregular fruits are around 6cm (2¼in) across. They are bright yellowish green with a red flush. The whitish flesh is soft, with a sweet, subacid flavour.

Type: Dessert.
Origin: Belgium, 1948.
Parentage: Unknown.
Flowering: Mid-season.

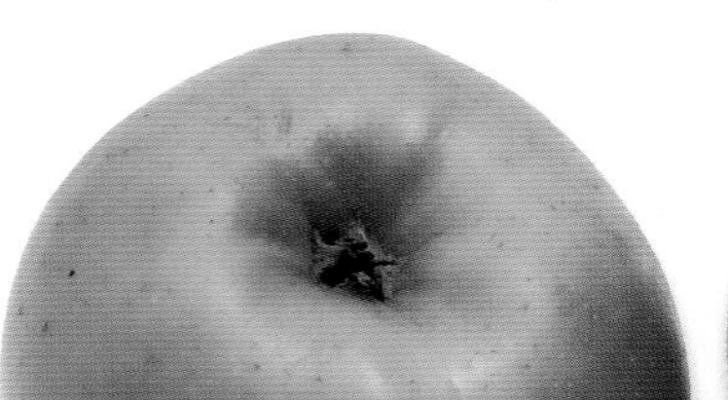

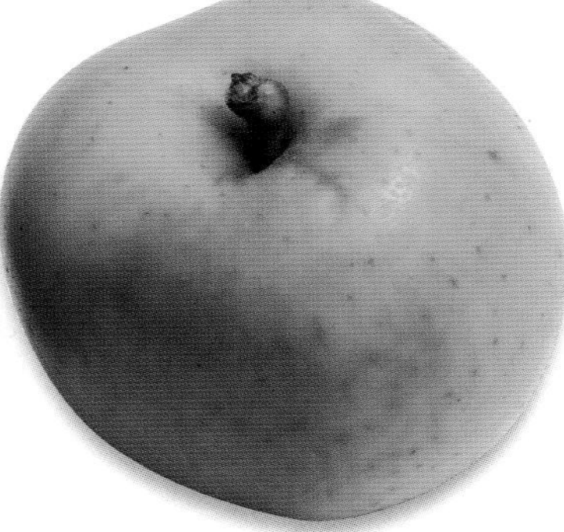

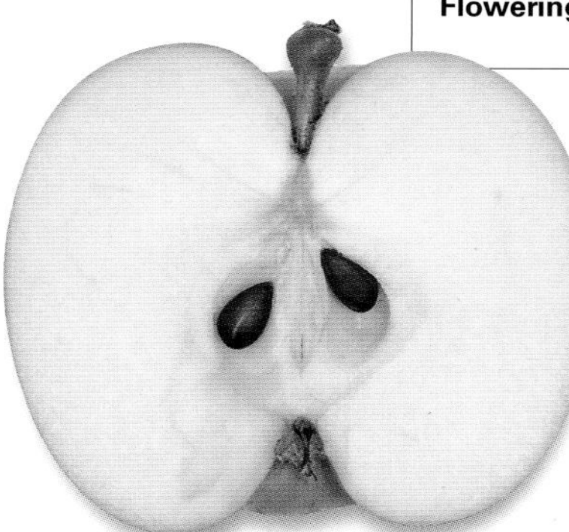

Note: Trees are vigorous.

Cockett's Red

The slightly flattened, somewhat irregular fruits are around 6cm (2¼in) across. They are light yellowish green with a usually even, bright red flush. The whitish flesh is firm and sweet in flavour, sometimes sharp when first picked.

Type: Dessert.
Origin: Wisbech, Cambridgeshire, UK, 1929.
Parentage: Unknown.
Flowering: Mid-season.
Other names: Marguerite Henrietta and One Bite.

Note: Fruits store well, the flavour becoming more mellow. This variety was popular for making toffee apples.

Cola

The distinctly conical fruits are around 5cm (2in) across or more. They are an even bright yellowish green in colour. The whitish flesh is coarse in texture with a subacid flavour.

Note: This variety is well suited to growing in a warm Mediterranean climate and does well in coastal situations.

Type: Dessert.
Origin: Italy, early 1900s.
Parentage: Unknown.
Flowering: Mid-season.

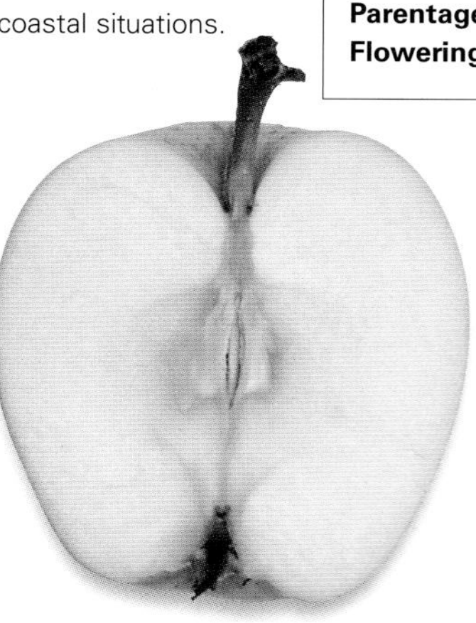

Coquette d'Auvergne

Type: Dessert.
Origin: France, 1947.
Parentage: Unknown.
Flowering: Late.

The rather conical fruits are around 6cm (2¼in) across. The skins are light yellow-green with a pronounced dark red flush. The whitish flesh is sometimes tinged orange immediately beneath the skin. It is soft and rather mealy in texture.

Note: Trees are weak-growing. Nowadays, this variety is rare in cultivation.

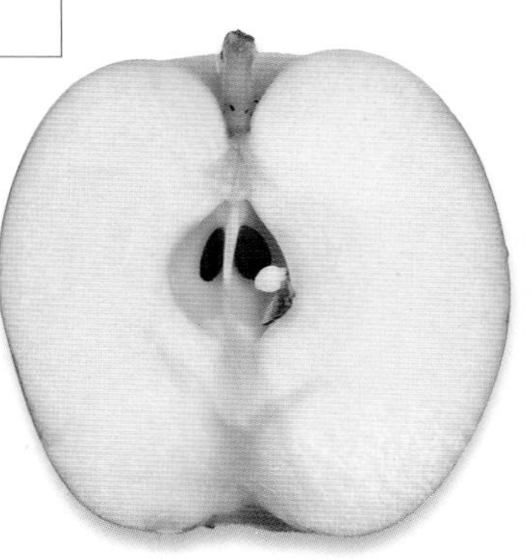

Cornish Aromatic

Type: Dessert.
Origin: Cornwall, UK, recorded 1813 but probably much older.
Parentage: Unknown.
Flowering: Late.
Other names: Aromatic, Aromatic Pippin, Aromatic Russet, Aromatique de Cornouailles, Cornwalliser Gewurzapfel and Siberian Russet.

The handsome fruits, around 6cm (2¼in) across, are rounded but may be knobbly. The red flush is broken by some spotting and russeting. The creamy white flesh is firm and rather dry. The flavour is rich, spicy and aromatic.

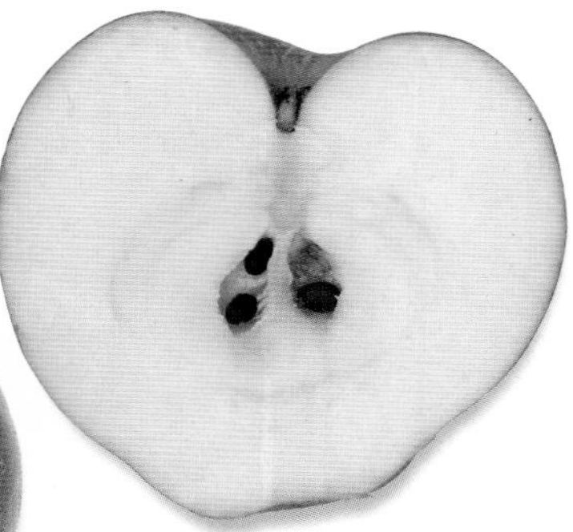

Note: This old English variety is thought to have been grown in Cornwall for many centuries. Trees bear freely.

Cornish Gilliflower

Type: Dessert.
Origin: Found in a cottage garden near Truro, Cornwall, UK, introduced in 1813.
Parentage: Unknown.
Flowering: Mid-season.
Other names: Calville d'Angleterre, Cornish Juli Flower, Cornwalliser Nelkenapfel, Gilliflower, Julie Flomer, July Flower, Kalvil angliiskii, Nelken Apfel, Pomme Regelans, Red Gilliflower and Regalan.

The irregular, bumpy fruits, up to around 8cm (3in) across, are dull yellow-green with a pinkish red flush with a webbing of rough russeting. The creamy white flesh is firm and rather dry with a sweet and rich aromatic flavour.

Note: This variety is tip-bearing, seldom setting abundant crops. It was an important variety in the 19th century. The flowers are clove scented. Fruits store well for three months or longer. For the best flavour, pick them as late as possible.

Above: Cornish Gilliflower is one of the most desirable of dessert apples, because of its rich flavour.

Cortland

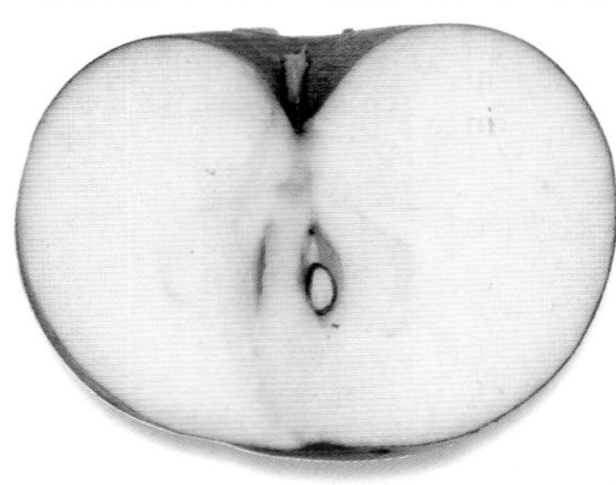

The fruits are flattened and rather irregular and can be up to 7cm (2¾in) across or more. They are bright greenish yellow with an uneven deep red flush. The white flesh is moderately juicy and slightly coarse in texture but with a sweet, refreshing flavour.

Type: Dessert.
Origin: New York State Agricultural Experiment Station, Geneva, USA, 1898.
Parentage: Ben Davis (female) x McIntosh (male).
Flowering: Early.
Other names: Cartland, Courtland and Courtlandt.

Note: The skins of this variety can be tough. Despite its early flowering, it crops well in cold areas. Cut fruits are slow to brown, so are suitable for use in fruit salads.

Cottenham Seedling

The fruits are irregularly rounded and around 7cm (2¾in) across. They are bright yellowish green with a pinkish red flush. The whitish flesh is firm, coarse-textured and juicy with a sharp, distinctly acid flavour.

Note: This is a fine cooking apple with very attractive blossom. Late flowering makes it useful for frost-prone areas.

Type: Culinary.
Origin: Cottenham, Cambridgeshire, UK, 1924.
Parentage: Dumelow's Seedling (female) x Unknown.
Flowering: Late.

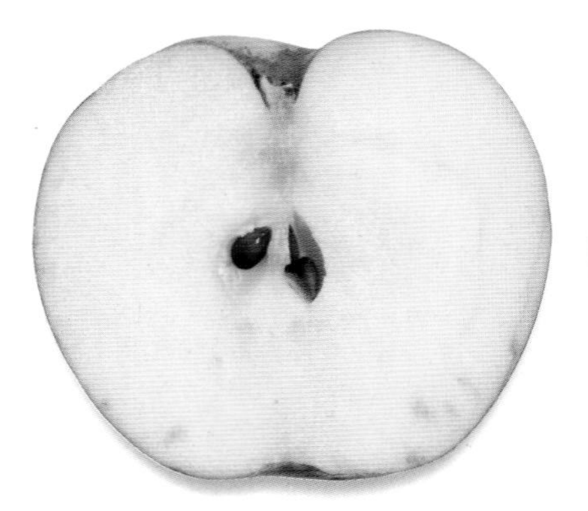

Court Pendu Plat

The flattened fruits, around 5cm (2in) across, are bright yellow with a bright orange and red flush and some spotting and striping. The creamy white flesh is very firm, fine-textured and juicy with a rich, sweet, slightly aromatic flavour.

Note: This excellent old variety possibly dates back to Roman times but it remains of more than just historic interest. It was called the 'wise apple' because it flowers late and hence escapes frost damage in cold areas. Fruits are also suitable for cooking.

Type: Dessert.
Origin: Europe, first described 1613 but believed to be much older.
Parentage: Unknown.
Flowering: Very late.
Other names: Belin, Belle de Senart, Capendu, Coriander Rose, Courte Queue, Garron's Apple, Kasapgel, Pomme de Berlin, Princesse Noble Zoete, Prudente, Reinette de Belges, Reinette de Portugal, Reinette Plate d'Hiver, Reinette Rose, Rode Korpendu, Roter Kurzstiel, Russian Apple, Wise Apple, Wollaton Pippin and Zlatousek Kratkostopkaty.

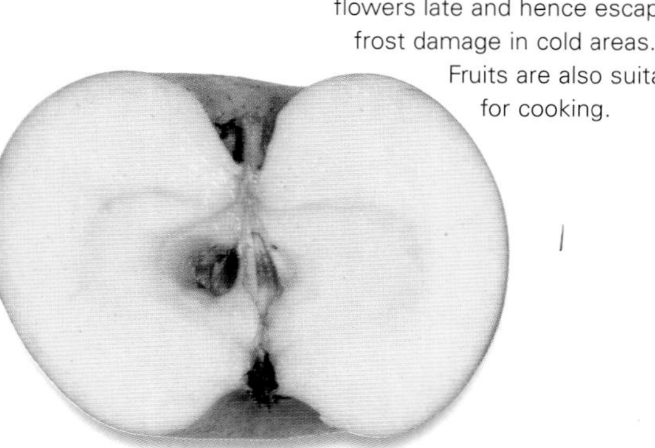

Cox's Orange Pippin

Type: Dessert.
Origin: Buckinghamshire, UK, 1825.
Parentage: ?Ribston Pippin (female) x Unknown.
Flowering: Mid-season.
Other names: Apelsinnyi renet, Cox Orangen Reinette, Cox's Pomeranzen Pepping, Coxova Reneta, Kemp's Orange, Koksa Pomaranczowa, Orange de Cox, Reinette Orange de Cox, Renet Coksa, Renet Cox Portocaliu, Reneta Coxa pomaranzowa, Russet Pippin and Verbesserte Muscat Reinette.

The rounded fruits are up to 5cm (2in) across, sometimes smaller. They are yellowish green with a red stripe and an orange flush. The creamy white flesh is crisp and juicy with an intensely aromatic flavour.

Note: This is sometimes regarded as the best of all English apples. It does best in warm, dry climates – elsewhere, it can be prone to scab and canker. The flavour improves in storage.

Crawley Beauty

Type: Dessert and culinary.
Origin: Crawley, Sussex, UK, *c.*1870, introduced 1906.
Parentage: Unknown.
Flowering: Very late.
Other names: Ratcliff Sargeant and Ratcliffe Sargeant.

The rounded fruits, around 6cm (2¼in) across, are bright green with a red flush and some spotting. The whitish flesh is slightly coarse in texture and rather dry. The flavour is sharp and slightly sweet.

Note: Late flowering makes this disease-resistant variety useful for areas where spring frosts are likely. Growth is vigorous and upright. It is apparently identical with the French variety Novelle France.

Crimson Queening

Type: Dessert.
Origin: England, first recorded 1831 but probably much older.
Parentage: Unknown.
Flowering: Mid-season.
Other names: Crimson Quoining, Herefordshire Queening, Quining, Red Queening, Scarlet Queening, Summer Queening and Summer Quoining.

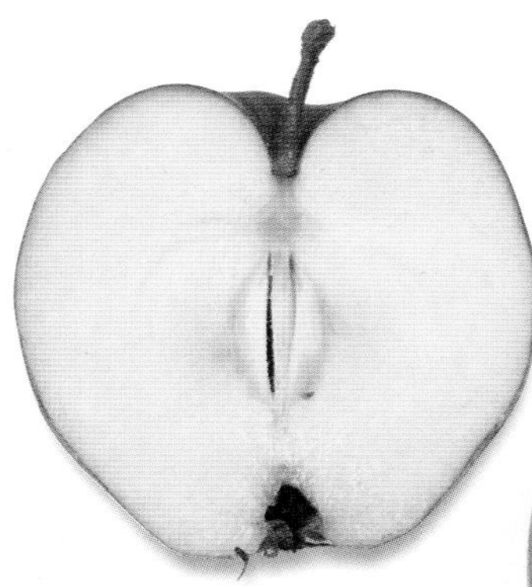

This old English variety is notable for the pointed shape of its fruits, which are around 5cm (2in) across. They are bright green with an even dark red flush. The creamy white flesh is soft and dryish, becoming mealy. The flavour is only moderately sweet.

Note: Trees are weak-growing. The fruits can be stored for three to four months.

Cusset Blanc

The flattened, slightly irregular fruits are around 6cm (2¼in) across. They are bright yellowish green with some flushing. The creamy white flesh is firm and coarse with a sharp flavour.

Note: Late flowering makes this a suitable variety for growing in cold areas.

Type: Dessert.
Origin: France, 1947.
Parentage: Unknown.
Flowering: Very late.

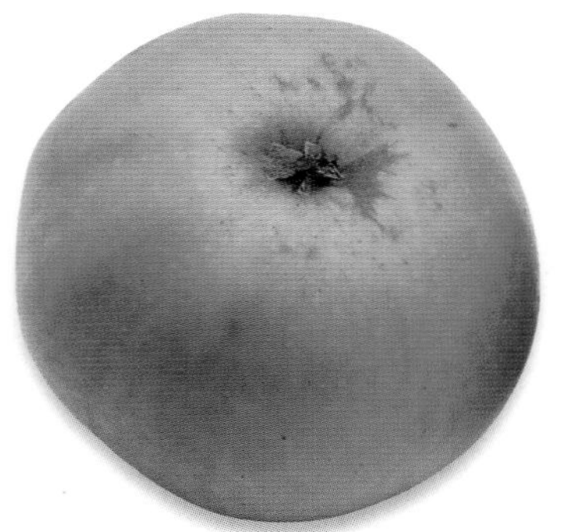

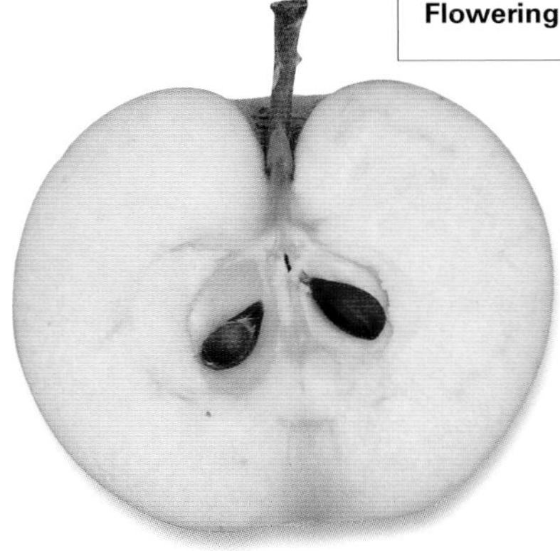

D'Arcy Spice

The oblong-shaped fruits, around 5cm (2in) across or more, are green with a crimson flush and pronounced russeting. The whitish flesh is firm, fine-textured and juicy with a characteristic spicy, aromatic flavour.

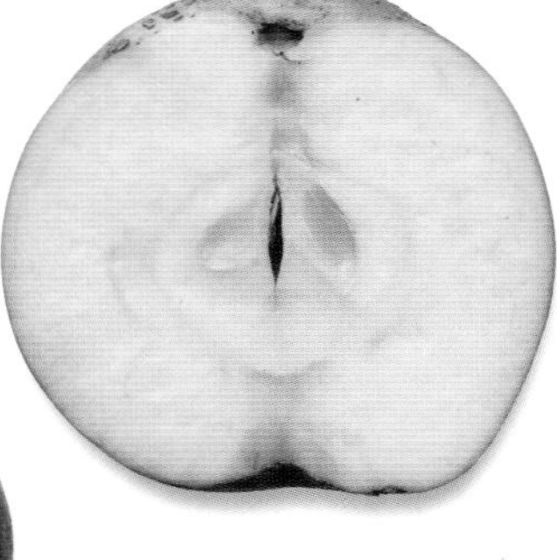

Note: The fruits of this excellent variety can be stored for four or five months, sometimes longer. Trees are moderately vigorous.

Type: Dessert.
Origin: Found in the garden of The Hall, Tolleshunt d'Arcy, Essex, UK, 1785 (but possibly older; introduced 1848 as Baddow Pippin).
Parentage: Unknown.
Flowering: Mid-season.
Other names: Baddow Pippin, Essex Spice, Pepin de Baddow, Spice, Spice Apple, Spring Ribston, Spring Ribston Pippin, Spring Ribstone and Winter Ribston.

Dalice

Above: Dalice is one of many dessert apples bred from Cox's Orange Pippin.

Note: Trees are moderately vigorous.

The rounded, very slightly conical fruits are around 6cm (2¼in) across. The skins are bright green flushed and marked with bright red. The creamy white flesh is coarse, soft and dry with a subacid to sweet, insipid flavour.

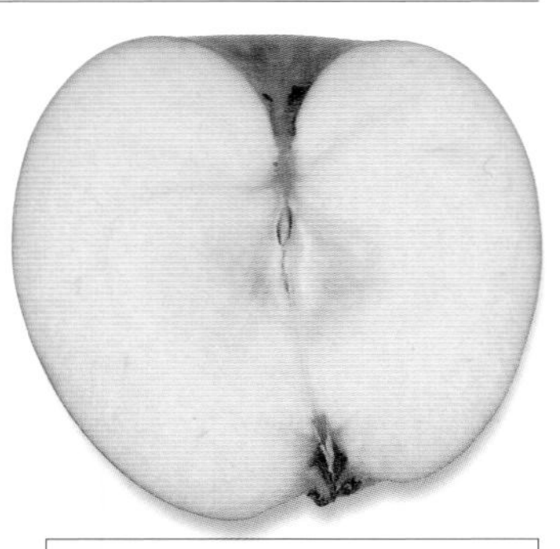

Type: Dessert.
Origin: Hastings, Sussex, UK, between 1933 and 1937.
Parentage: Cox's Orange Pippin (female) x Unknown.
Flowering: Mid-season.

Dark Red Staymared

Type: Dessert.
Origin: Barber County, VA, USA, introduced 1927.
Parentage: Winesap (female) x Unknown.
Flowering: Mid-season.
Other name: Dark-red Staymared.

The rounded to somewhat conical fruits, around 6cm (2¼in) across, are bright-green with a strong dark red flush and some striping. The pale yellow flesh is juicy with a light aromatic flavour.

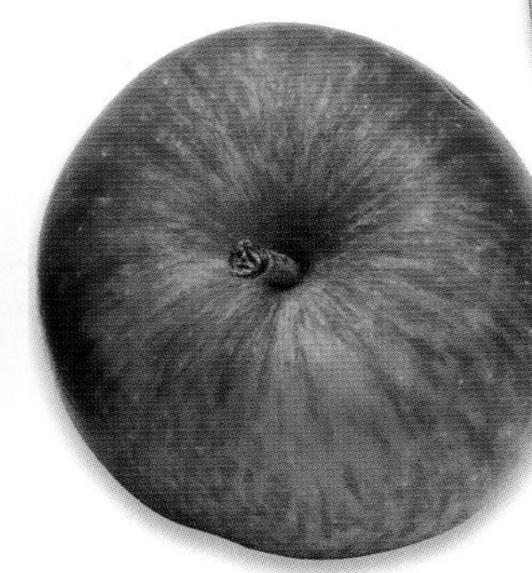
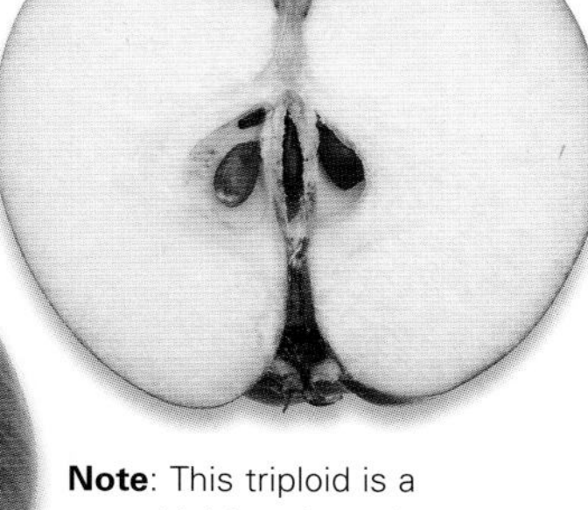

Note: This triploid is a more highly coloured sport of Stayman's Winesap. They are bright green with a heavy, deep red flush.

Dawn

Type: Dessert.
Origin: ?Ware Park Gardens, Hertfordshire, UK, 1940.
Parentage: Unknown.
Flowering: Mid-season.

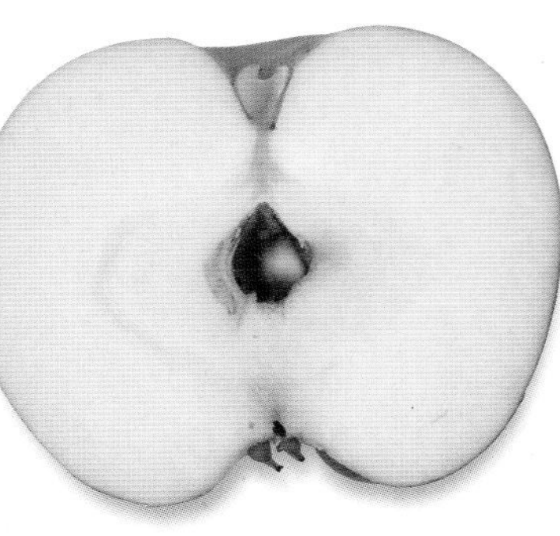

The rounded (but sometimes slightly pointed) fruits are around 6cm (2¼in) across and bright yellow-green with a bright red flush. The crisp white flesh has a sweet, sharp flavour that is somewhat reminiscent of raspberries.

Note: This variety is self-sterile. Trees are moderately vigorous.

Left: A suitable pollination partner nearby will ensure good fruiting.

de Flandre

Type: Dessert.
Origin: France, 1876.
Parentage: Unknown.
Flowering: Late.
Other name: Figue.

The somewhat flattened fruits, around 6cm (2¼in) across, are bright green with a strong red flush. The creamy white flesh is firm and coarse with a slightly sweet, subacid and aromatic flavour.

Note: This variety is believed to show some resistance to blackspot. Trees are moderately vigorous.

De Vendue l'Eveque

The rounded fruits, around 6cm (2¼in) across, are an even bright green, lightly speckled with pale yellowish white dots. The creamy white flesh is crisp with a tart flavour.

Type: Culinary.
Origin: France, 1948.
Parentage: Unknown.
Flowering: Late.

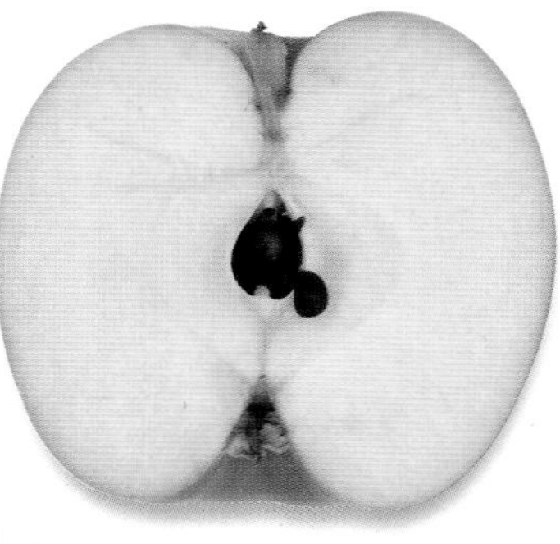

Note: This variety is commonly used in France as a cider apple. Late flowering makes it suitable for growing in cold areas.

Right: This variety is grown in the Aube department of France.

Delcorf

Type: Dessert.
Origin: Malicorne, France, 1956 and introduced into commerce 1976.
Parentage: Stark Jonagrimes (female) x Golden Delicious (male).
Flowering: Mid-season.
Other names: Ambassy Dalili Delbarestivale, Delcorf Estivale and Estivale Monidel.

Note: Trees are of average vigour. Fruits bruise easily.

The rounded to slightly oblong fruits, around 7cm (2¾in) across, are bright yellow-green with a strong pinkish or orange-red flush and some flecking. The creamy white flesh is fairly crisp and juicy and very sweet, but with a hint of sharpness.

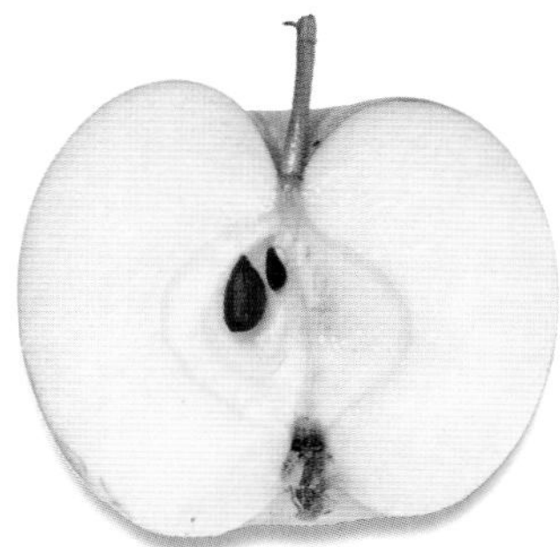

Above: Delcorf is grown commercially in both France and England.

Delicious

The irregular fruits, around 6cm (2¼in) across or more, are yellow-green with a strong dark red flush. The whitish flesh is very firm, very sweet and juicy with a highly aromatic flavour.

Type: Dessert.
Origin: Peru, Iowa, USA, *c.*1880 (introduced 1895).
Parentage: Unknown.
Flowering: Mid-season.
Other names: Cervena prevazhodna, Delicious rosso, Hawkeye, Piros Delicious, Prevoshodnoe krasnoe and Stark Delicious.

Note: Fruits develop their best flavour in warm areas. The skin can be very tough.

Delgollune

The conical fruits, around 6cm (2¼in) across, are bright greenish yellow with a pinkish red flush. The whitish flesh is crisp and juicy with a sweet, lightly aromatic flavour.

Type: Dessert.
Origin: Delbard Nurseries, France, 1962.
Parentage: Golden Delicious (female) x Lundbytorp (male).
Flowering: Very late.
Other name: Delbard Jubilee.

Note: Late flowering makes this variety suitable for growing in areas with late spring frosts.

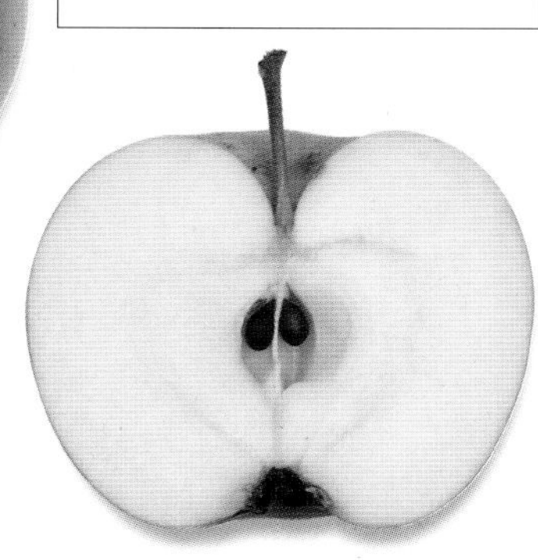

Left: Delgollune is a popular commercial variety in France and has been used in breeding programmes.

Delnimb

Type: Dessert.
Origin: Malicorne, France, 1960s.
Parentage: Maigold (female) x Grive Rouge (male).
Flowering: Very late.

The rounded fruits, around 6cm (2¼in) across, are an even bright yellow-green. The creamy white flesh is rather coarse but juicy with a fairly rich, sweet, subacid flavour.

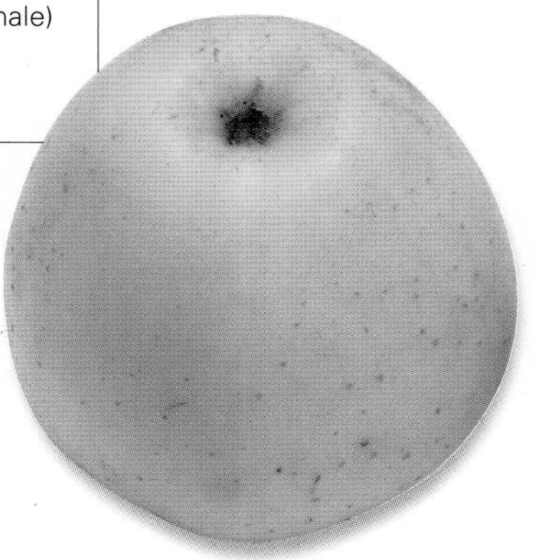

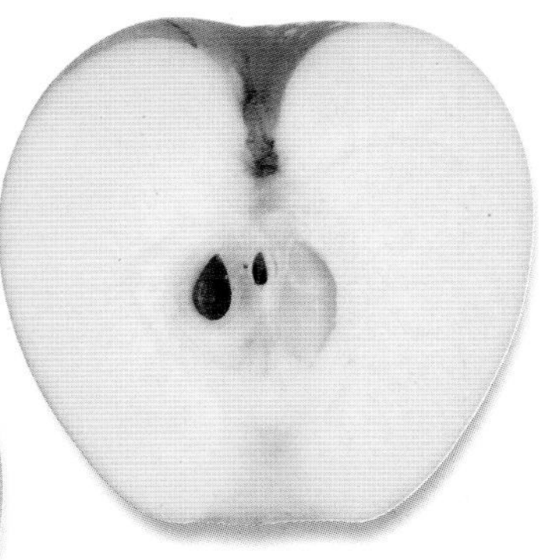

Note: The late flowering of this variety makes it suitable for growing in areas that are prone to late spring frosts.

Devonshire Quarrenden

The flattened, slightly irregular fruits, around 7cm (2¾in) across, are bright yellow-green with a strong dark purplish red flush. The creamy white flesh, sometimes stained red, is sweet, crisp and juicy with a distinctive aromatic flavour.

Type: Dessert.
Origin: Believed to be Devon, UK, but possibly originally France, first recorded 1678.
Parentage: Unknown.
Flowering: Early.
Other names: Annat Scarlet, Englischer Scharlach Peppin, Morgenrotapfel, Pepin alyi, Pippin Scarlet, Pomme Impériale, Quarrendon du Comité de Devon, Quarrington, Red Quarrenden, Sack, Scarlet Pippin, Scharlach Pepping, Sharlakhovyi pepin and Tsyganka.

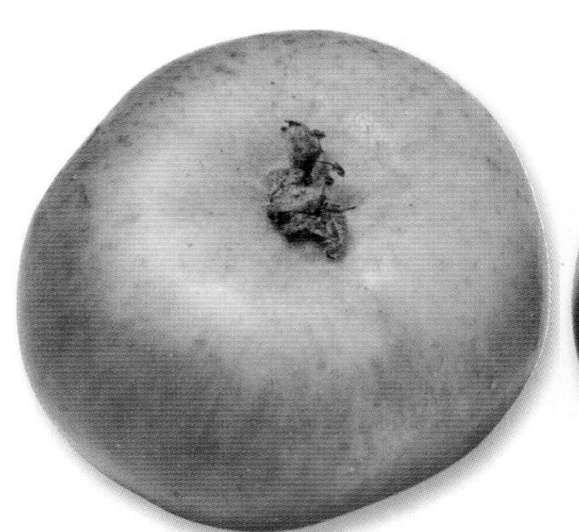

Note: This variety is a triploid. Trees are hardy and crop freely.

Diamond Jubilee

Type: Dessert and culinary.
Origin: Kent, UK, 1889.
Parentage: Unknown.
Flowering: Mid-season.

The rounded fruits, around 7cm (2¾in) across, are bright green or greenish yellow. The whitish flesh is firm and crisp with a subacid and slightly aromatic flavour.

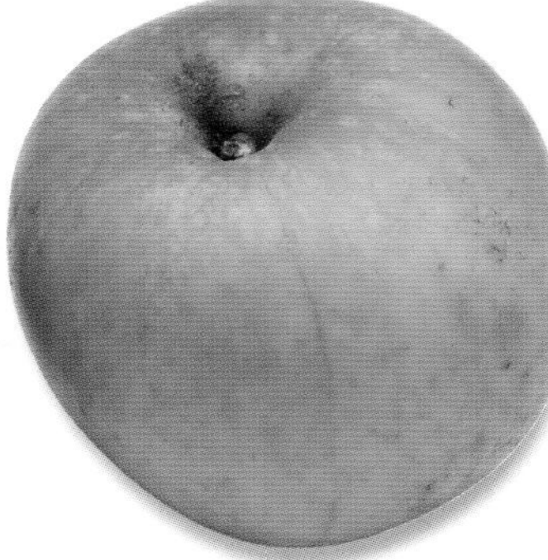

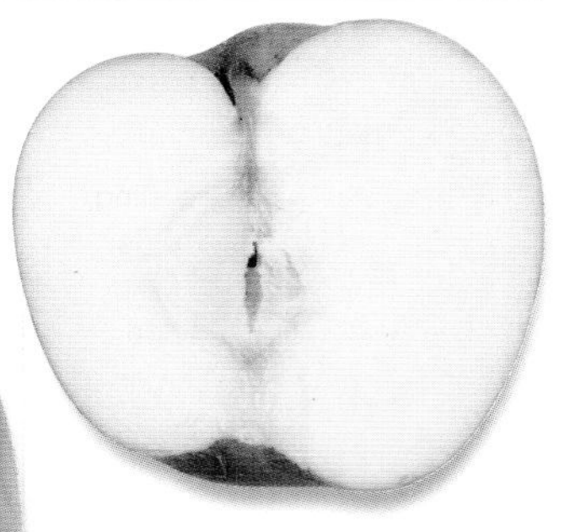

Far left: Skins of this variety show an even colouring.

Note: This old variety was named to commemorate Queen Victoria's Diamond Jubilee.

Diana

The rounded fruits, around 7cm (2¾in) across, are bright greenish yellow with a strong dark red flush. The whitish flesh is sweet, soft and juicy.

Note: The flavour is similar to McIntosh. Trees are moderately vigorous.

Type: Dessert.
Origin: Former Yugoslavia, 1967.
Parentage: Unknown.
Flowering: Mid-season.

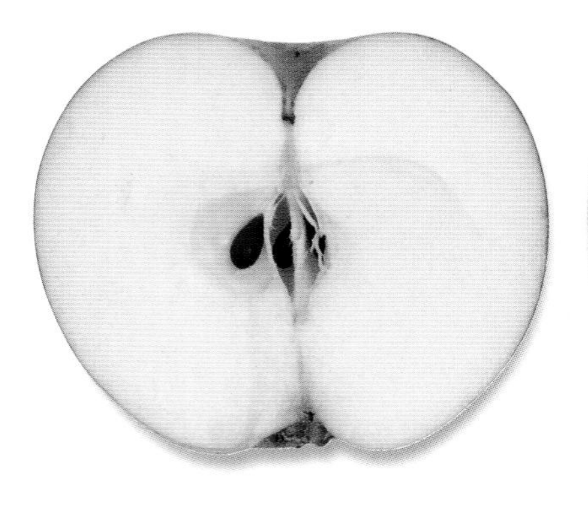

Dillington Beauty

The slightly flattened, rather irregular fruits, around 7cm (2¾in) across, are dull yellowish green with some flushing and spotting. The whitish flesh is soft and rather coarse with a subacid flavour.

Type: Dessert.
Origin: New Zealand, 1872.
Parentage: Unknown.
Flowering: Mid-season.

Note: Formerly popular as a garden variety in New Zealand, Dillington Beauty has become rare in cultivation. Trees are moderately vigorous.

Directeur van de Plassche

Type: Dessert.
Origin: Institute for Horticultural Plant Breeding, Wageningen, Netherlands, 1935.
Parentage: Cox's Orange Pippin (female) x Jonathan (male).
Flowering: Mid-season.

The rounded fruits, around 6cm (2¼in) across, are bright greenish yellow with a bright red flush. The creamy white flesh is juicy with a slightly subacid flavour.

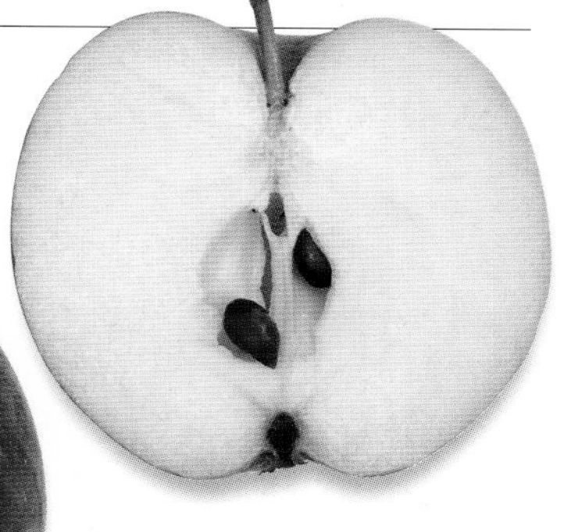

Note: Trees are moderately vigorous.

Dorée de Tournai

Type: Dessert.
Origin: Tournai, Belgium, 1817.
Parentage: Unknown.
Flowering: Early.
Other names: Doree de Tournay, Gold Apfel von Tournay and la Dorade.

The irregular, somewhat square fruits, around 7cm (2¾in) across, are dull yellowish green. The flesh is firm and crisp with a sweet, subacid, rich, aromatic flavour.

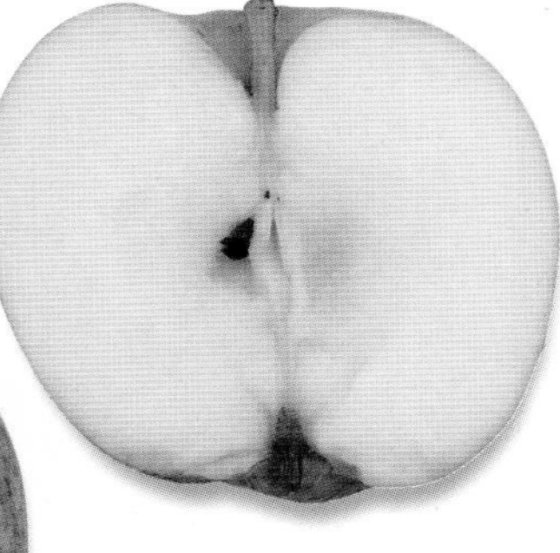

Note: This variety is nowadays rare in cultivation outside historic gardens.

Dorsett Golden

The irregular fruits, around 7cm (2¾in) across, are yellowish green with a strong bright red flush. The creamy white flesh is juicy with a sweet, light, aromatic flavour.

Note: This variety is an excellent choice for warm and coastal areas. The young trees bear freely.

Type: Dessert.
Origin: Nassau, New Providence Islands, 1953.
Parentage: Golden Delicious (female) x Unknown.
Flowering: Late.

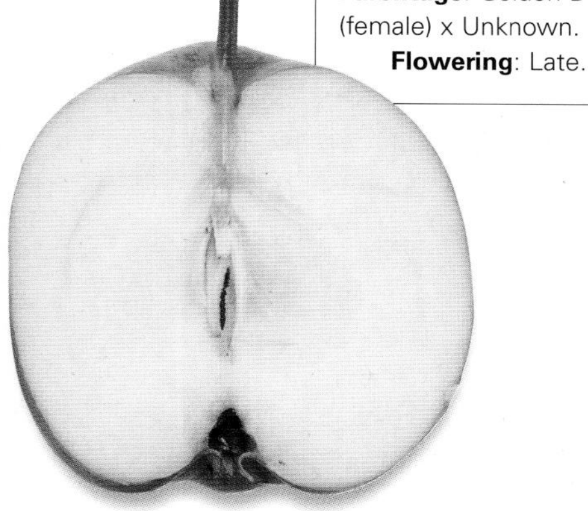

Double-Red Baldwin

Type: Dessert.
Origin: Salisbury, New Hampshire, USA, 1924.
Parentage: Unknown.
Flowering: Early.

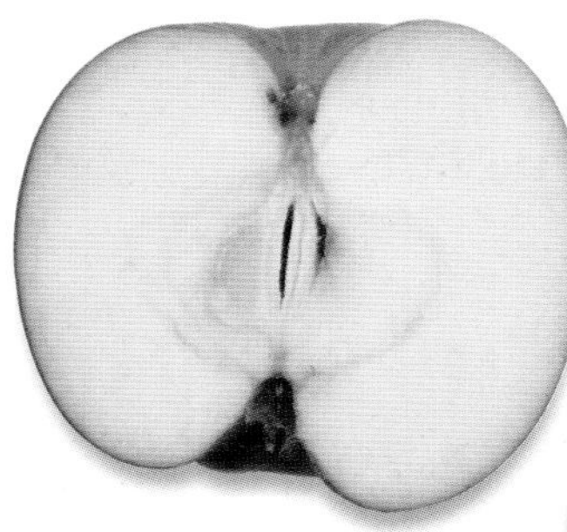

The flattened, irregular fruits, around 6cm (2¼in) across, are bright yellow-green with a strong dark red flush. The yellowish flesh is coarse-textured and juicy with a sweet, subacid flavour.

Note: This variety is a more brightly coloured sport of Baldwin.

Left: The skins of this variety ripen to a very deep red on the side nearest the sun.

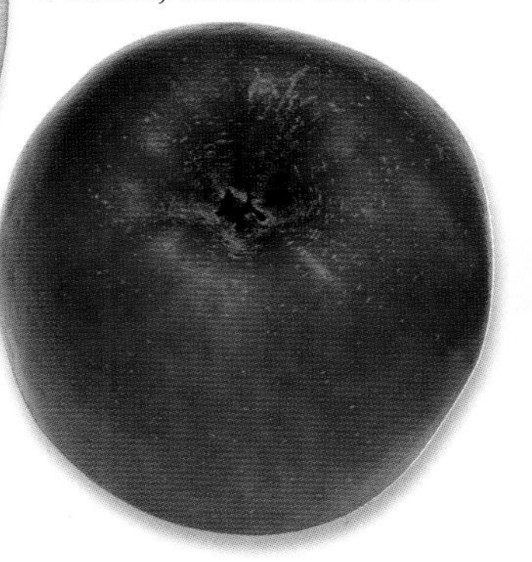

Dugamel

The rounded, slightly knobbly fruits, around 6cm (2¼in) across, are bright green with a strong and even dark red flush. The whitish flesh is firm and juicy with a pleasant, faintly aromatic flavour.

Type: Dessert.
Origin: France, date unknown.
Parentage: Unknown.
Flowering: Mid-season.

Note: This variety is a more highly coloured clone of Melrose.

Duke of Devonshire

The flattened fruits, around 5cm (2in) across, are bright yellow-green and heavily russeted. The creamy white flesh is firm, fine-textured and rather dry with a rich, nutty, intense sweet-sharp flavour.

Note: Fruits store well, developing their best flavour after one to two months. Trees are moderately vigorous.

Type: Dessert.
Origin: Holker Hall, Lancashire, UK, 1835 (introduced 1875).
Parentage: Unknown.
Flowering: Mid-season.
Other names: Devonshire Duke, Duc de Devonshire, Herzog von Devonshire, Holker and Holker Pippin.

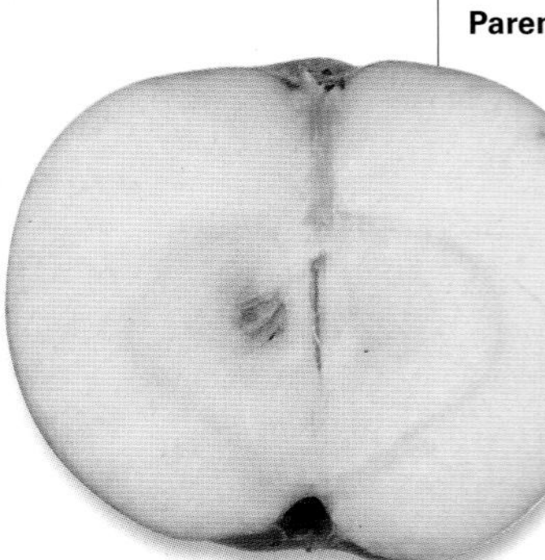

Dunn's Seedling

Type: Dessert.
Origin: Kew, Melbourne, Australia, first recorded 1890.
Parentage: Unknown.
Flowering: Early.
Other names: Chenimuri, Dunn's Favourite, Dunns, Monroe's Favourite, Munroe's Favourite, Ohenimuri and Ohinemuri.

The rounded, slightly bumpy fruits, around 6cm (2¼in) across, are bright yellow-green with some flushing. The creamy white flesh is crisp and hard with a sweet, subacid flavour.

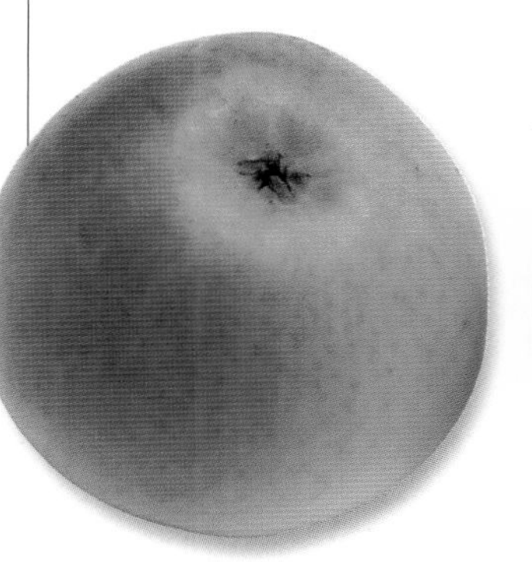

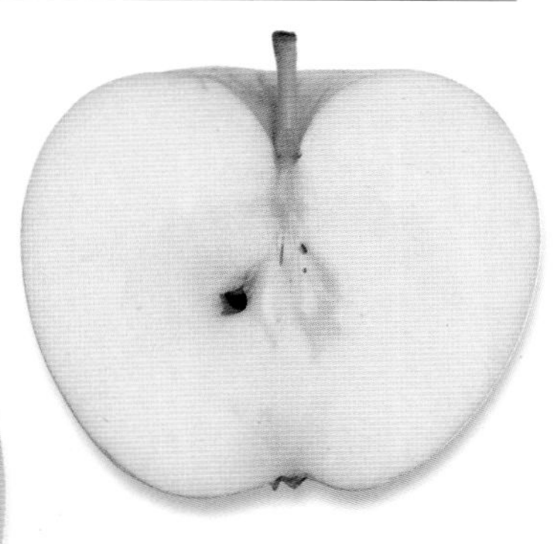

Note: This variety is important historically as one of the first to be grown widely in Australia. Trees are moderately vigorous.

Dutch Mignonne

Type: Dessert and culinary.
Origin: ?Netherlands, before 1771.
Parentage: Unknown.
Flowering: Early.
Other names: Belle Reinette de Caux, Casseler Reinette, Christ's Golden Reinette, Contor, Copmanshorpe Crab, Copmanshorpe Russet, Craft Angry, de Laak, Dutch Minion, Grosse-Reinette Rouge Tiquetee, Hollandische Goldreinette, Paternoster, Rawle's Reinette, Reinette Imperatrice, Stettiner Pepping, Thorpe Grabe and Vermillon d'Andalousie.

The flattened fruits, around 6cm (2¼in) across, are bright green with a light flush. The creamy yellow flesh is firm and juicy with a slightly acid, not very sweet, faintly aromatic flavour.

Note: Trees bear freely and are suitable for training as espaliers. Fruits store well.

Right: This old variety readily forms spur systems that carry regularly shaped fruits.

Easter Orange

Type: Dessert.
Origin: Winchester, Hampshire, UK, 1897.
Parentage: Unknown.
Flowering: Mid-season.

Note: The fruits are also suitable for cooking.

Left: When ripe, the fruits of this variety have a very intense flavour.

The rounded fruits, around 5cm (2in) across, are yellow-green with a strong red flush. The creamy white flesh is crisp and firm with a sweet and aromatic flavour.

Egremont Russet

The slightly flattened fruits, around 6cm (2¼in) across, are yellow-green with heavy russeting. The creamy white flesh is firm, fine textured and rather dry, with a very distinctive rich, nutty flavour.

Note: This is probably the best-known and most popular of the russet apples. Trees, which are hardy and compact, are partially self-fertile.

Type: Dessert.
Origin: ?England, first recorded 1872.
Parentage: Unknown.
Flowering: Early.
Other name: Egremont.

Above: Egremont Russet is a classic English russet apple from the Victorian era.

Elan

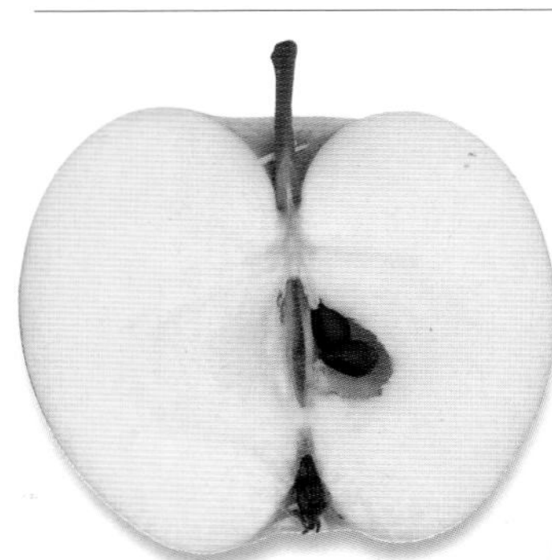

Note: This variety is tip-bearing. Trees are moderately vigorous.

The rounded fruits, around 7cm (2¾in) across, are greenish yellow with red flushing and striping. The creamy white flesh is crisp, sweet and juicy.

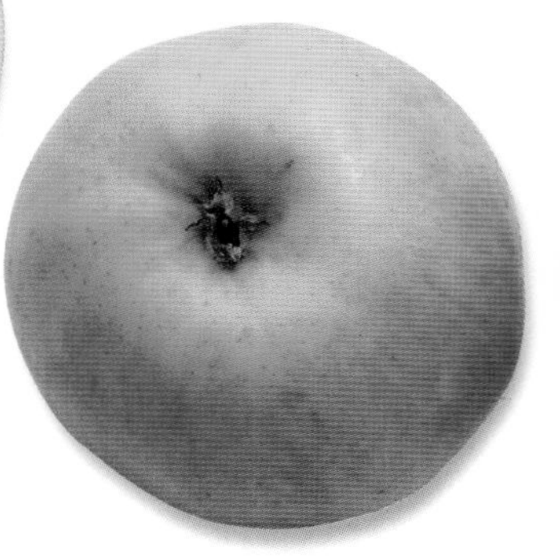

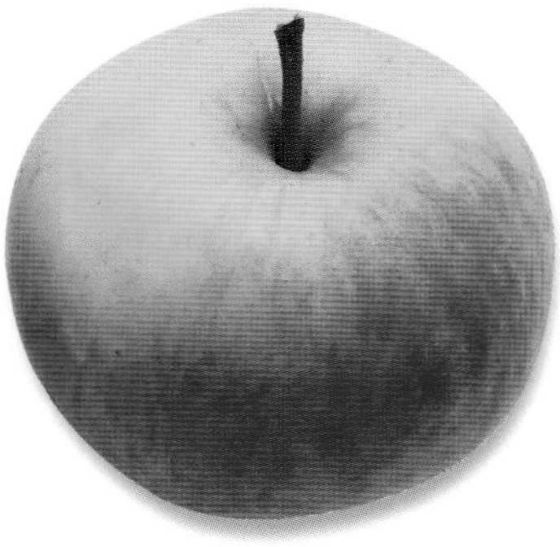

Type: Dessert.
Origin: IVT, Wageningen, Netherlands, before 1983.
Parentage: Golden Delicious (female) x James Grieve (male).
Flowering: Very early.

Elise Rathke

The rounded fruits, around 5cm (2in) across, are yellowish green with a red flush and striping. The creamy white flesh is fine and soft with a subacid and moderately sweet and spicy flavour.

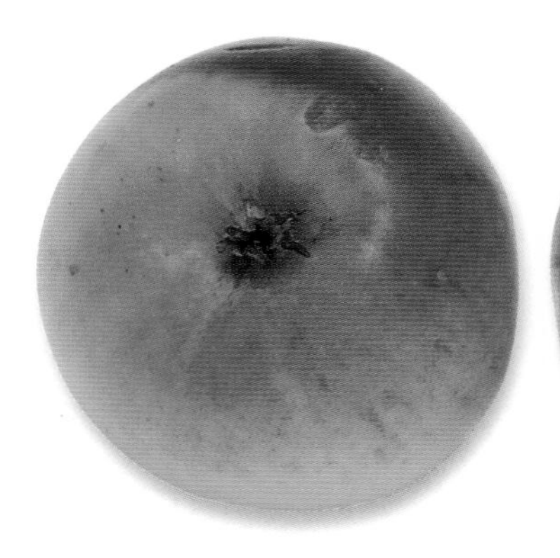

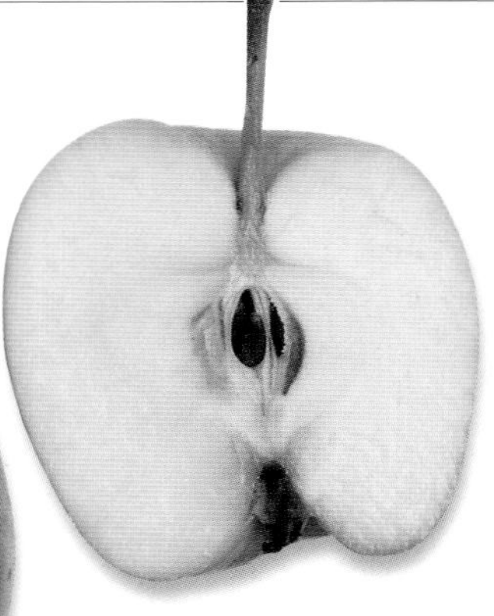

Note: This variety has a drooping habit and is good in a confined space.

Type: Dessert.
Origin: Pravst, nr Gdansk, Poland, or Elbinge, Germany, first recorded 1884.
Parentage: Unknown.
Flowering: Mid-season.
Other names: Elisa Rathké, Elisa Rathke, Elisa Ratk, Eliza Ratke and Rote Reinette.

Ellison's Orange

Type: Dessert.
Origin: Bracebridge and at Hartsholme Hall, Lincolnshire, UK, first recorded 1904 and introduced 1911.
Parentage: Cox's Orange Pippin (female) x Calville Blanc (male) (probably Calville Blanc d'Hiver).
Flowering: Mid-season.

The rounded to slightly conical fruits, around 6cm (2¼in) across, are yellow-green with a strong dark red flush and striping. The creamy white flesh is soft and juicy, like a pear, with a rich, strong aniseed flavour.

Note: The fruits are not suitable for long-term storing. The flavour is considered one of the most complex of all dessert apples. Trees are generally disease resistant and easy to grow.

Above: *This variety is one of the best offspring of Cox's Orange Pippin.*

Elmore Pippin

The bright greenish yellow, rounded fruits, around 5cm (2in) across, show some spotting. The creamy white flesh is firm with an intense, sweet-sharp flavour.

Note: Fruits can be stored for up to five months. Trees are moderately vigorous.

Type: Dessert.
Origin: ?UK, before 1949.
Parentage: Unknown.
Flowering: Mid-season.

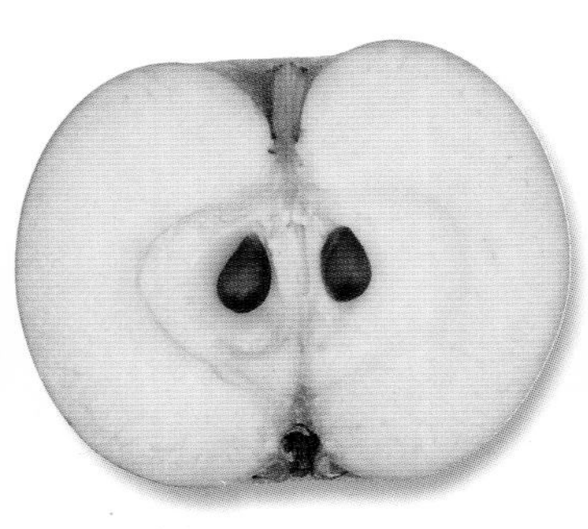

Right: *The fruits of Elmore Pippin stay on the tree until late autumn.*

Eri Zagarra

Type: Culinary.
Origin: France (Basque Country), 1973.
Parentage: Unknown.
Flowering: Mid-season.

The slightly flattened fruits, around 6cm (2¼in) across, are greenish yellow with a red flush and some russeting. The creamy white flesh is tart in flavour.

Note: Fruits cook well.

Left: *Eri Zagarra is an excellent cooking apple but somewhat rare in cultivation.*

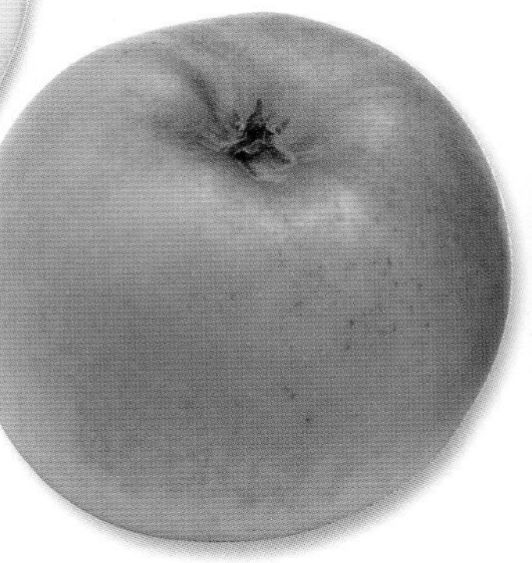

Esopus Spitzenburg

The somewhat conical fruits, around 6cm (2¼in) across, are bright greenish-yellow with a strong red flush. The whitish flesh is crisp and tender with a rich aromatic flavour.

Note: This old American apple was widely grown in the USA in the 19th century and was the traditional choice for apple pies. It does best in warm areas.

Type: Dessert.
Origin: Esopus, Ulster County, NY, USA, before 1790.
Parentage: Unknown.
Flowering: Mid-season.
Other names: Aesopus Spitzenberg, Aesopus Spitzenburg, Aesopus Spitzenburgh, Esopus, Esopus Spitzenberg, Esopus Spitzenburgh, Ezop Spitzenburg, Spitzenberg, Spitzenburg, True Spitzenberg and True Spitzenburg.

Etlins Reinette

The fruits are irregular in shape and can be up to around 7cm (2¾in) across. They are bright yellow-green with some spotting. The creamy white flesh is fine and crisp with a sweet, slightly subacid, aromatic flavour.

Note: This variety commemorates a noted German pomologist. Trees are very vigorous.

Type: Dessert.
Origin: Landenberg estate, Germany, 1866.
Parentage: Unknown.
Flowering: Early.
Other names: Etlin's Reinette, Ettlin's Reinette, Reinette d'Etlin and Reinette Ettlin's.

Fara Nume

The slightly conical fruits, around 6cm (2¼in) across, are dull yellow-green with some spotting. The creamy white flesh is soft and fine with a sweet, subacid flavour.

Type: Dessert.
Origin: Romania, 1948.
Parentage: Unknown.
Flowering: Mid-season.

Note: Fara nume is a Romanian phrase that translates as 'without a name'.

Right: The fine, speckled russeting on the fruits is characteristic of this variety.

Faversham Creek

Type: Culinary.
Origin: A seedling found growing in salt water in Faversham Creek, Kent, UK, 1970s.
Parentage: Unknown.
Flowering: Mid-season.

The slightly flattened fruits, around 6cm (2¼in) across, are bright yellow-green with a pinkish to orange-red flush and some striping. The yellowish flesh is coarse and dry with an acid flavour.

Note: This variety is very rare in cultivation. Fruits cook to a creamy consistency with very good flavour.

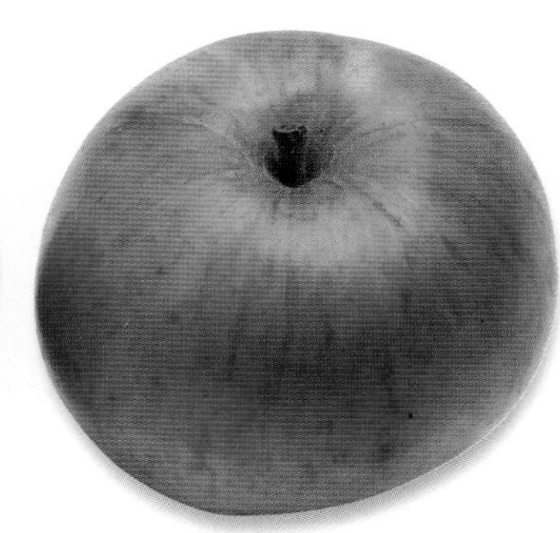

Fenouillet de Ribours

Type: Dessert
Origin: La Rouairie garden, Maine-et-Loire, France, first fruited 1840.
Parentage: Unknown.
Flowering: Mid-season.

The irregular, sometimes knobbly or ribbed fruits, up to around 7cm (2¾in) across, are bright green with a red flush and some greyish or bronze russeting and some white spotting. The white flesh is fine in texture with a sweet, subacid, aniseed-perfumed flavour.

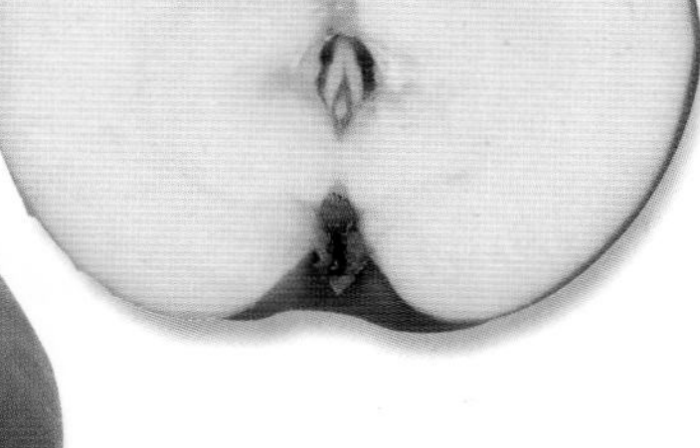
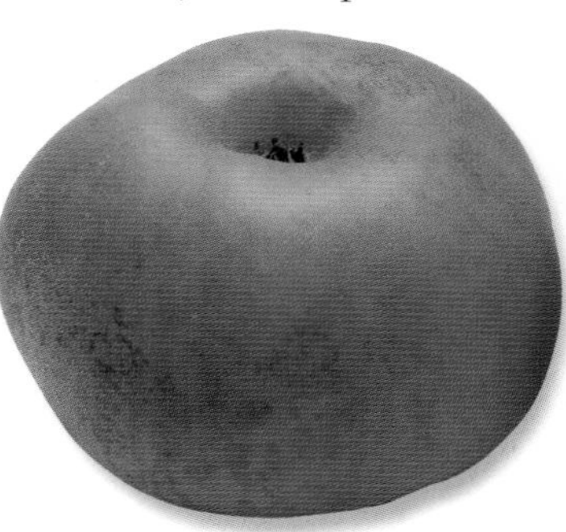
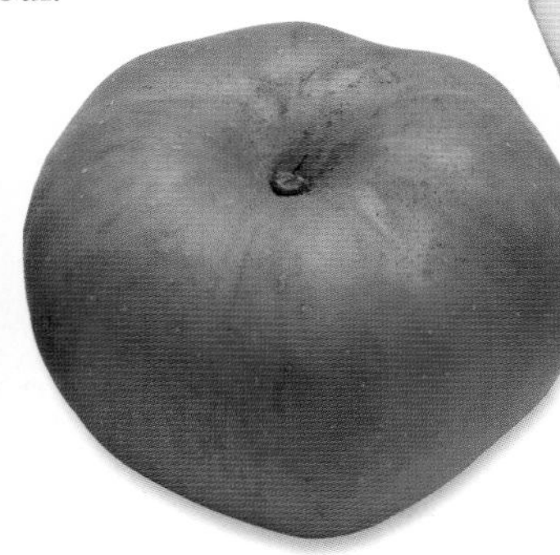

Note: The fruits are late to ripen on the tree. They are prized for their unusual flavour.

Feuillemorte

Type: Dessert.
Origin: France, 1948.
Parentage: Unknown.
Flowering: Very late.
Other name: Feuille Morte.

The somewhat flattened fruits are bright yellow-green with a strong red flush and striping. The white flesh is firm and fine with a subacid and distinctive flavour.

Note: This is one of the latest varieties to flower.

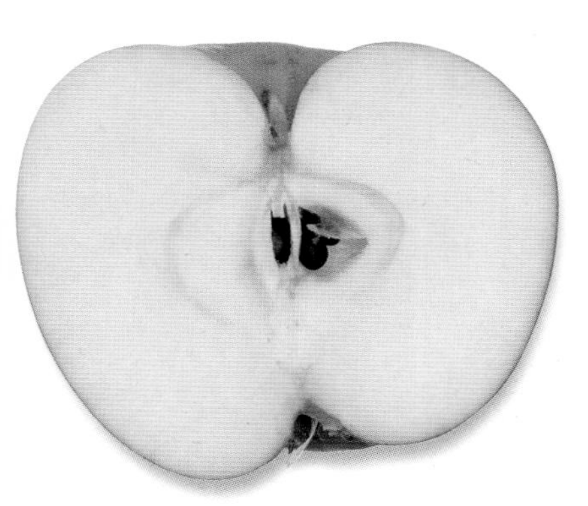

Finasso

The flattened fruits, around 6cm (2¼in) across, are yellowish green with a strong red flush. The whitish flesh is firm and fine with a slightly subacid flavour.

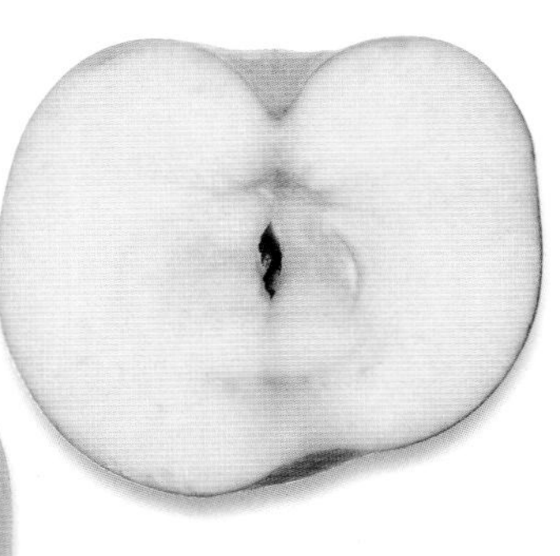

Type: Dessert.
Origin: France, 1949.
Parentage: Unknown.
Flowering: Late.

Note: Late flowering makes this a suitable variety for frost-prone areas.

Right: This moderately vigorous variety crops well.

Fireside

Type: Dessert.
Origin: Minnesota Agricultural Experiment Station, Excelsior, USA, 1917 (introduced commercially 1943).
Parentage: McIntosh x Longfield.
Flowering: Early.

The rounded fruits, around 6cm (2¼in) across, are bright green with a red flush and striping. The white flesh is firm and crisp with a sweet, subacid, aromatic flavour.

Note: Trees are vigorous, hardy and resistant to cedar apple rust. The fruits store well.

Firmgold

The conical, uneven fruits, around 6cm (2¼in) across, are greenish yellow with darker spotting. The flesh is firm, crisp and juicy with a sweet flavour.

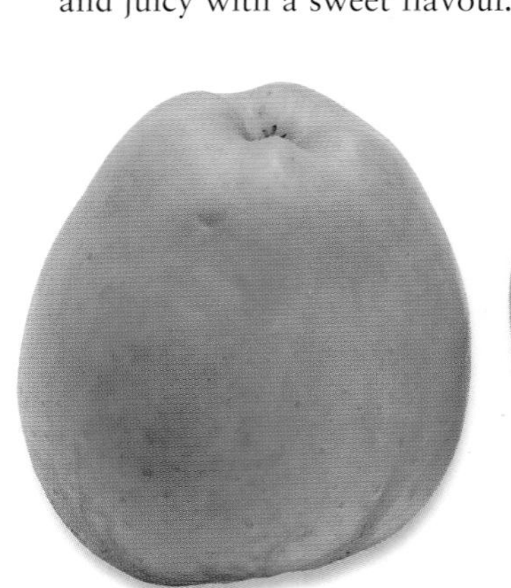

Type: Dessert.
Origin: Zillah, WA, USA, date not recorded.
Parentage: A chance seedling found growing among some Starkspur Golden Delicious and Starkrimson Red Delicious trees.
Flowering: Early.

Note: Trees are prone to biennial bearing if not thinned correctly.

Florina

Type: Dessert.
Origin: Station de Recherches d'Arboriculture Fruitière, Angers, France, date not recorded.
Parentage: Complex, involving Rome Beauty, Golden Delicious, Starking, Simpsons Giant Limb and Jonathan.
Flowering: Early.

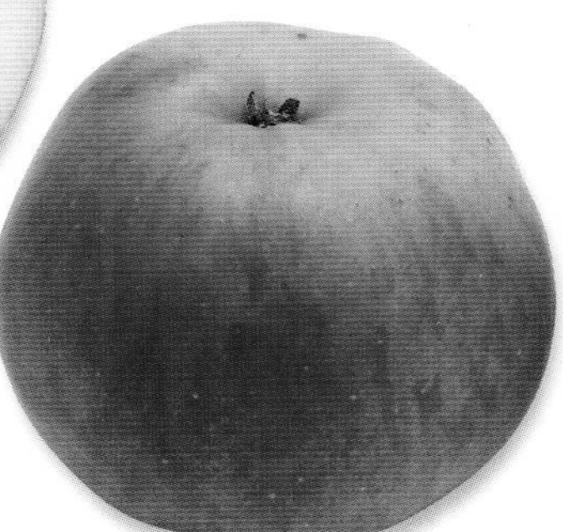

Note: This tip-bearing variety is resistant to scab.

The rounded fruits, around 6cm (2¼in) across, are bright yellow-green with a strong red flush and some spotting. The whitish flesh is rather tart and acid in flavour.

Fon's Spring

Type: Dessert.
Origin: Milbury Heath, Falfield, Gloucestershire, UK, 1948.
Parentage: John Standish (female) x Cox's Orange Pippin (male).
Flowering: Late.
Other name: Eden.

The slightly flattened fruits, around 6cm (2¼in) across, are bright yellow-green with a strong red flush. The white flesh is firm and fairly juicy with a sweet, subacid flavour, rather like Cox's Orange Pippin.

Note: Late flowering makes this a suitable variety for growing in cold districts where late frosts are common.

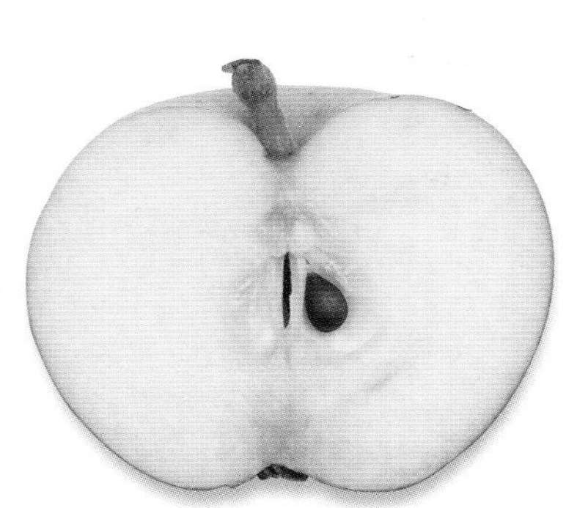

Forge

Type: Dessert.
Origin: East Grinstead, Sussex, before 1851.
Parentage: Unknown.
Flowering: Mid-season.
Other names: Der Schmiedeapfel, Forge Apple, Schmiede Apfel, Schmiedeapfel and Sussex Forge.

Note: The fruits are also suitable for cooking. Trees bear freely.

The rounded, rather uneven fruits, around 6cm (2¼in) across, are bright green with a red flush. The white flesh is crisp and very juicy with a pleasant, aromatic but somewhat sharp flavour.

Fortosh

The conical, unevenly shaped fruits, around 7cm (2¾in) across, are bright green with a red flush and streaks. The pink-tinged white flesh is soft and juicy with a moderately sweet, slightly acid flavour.

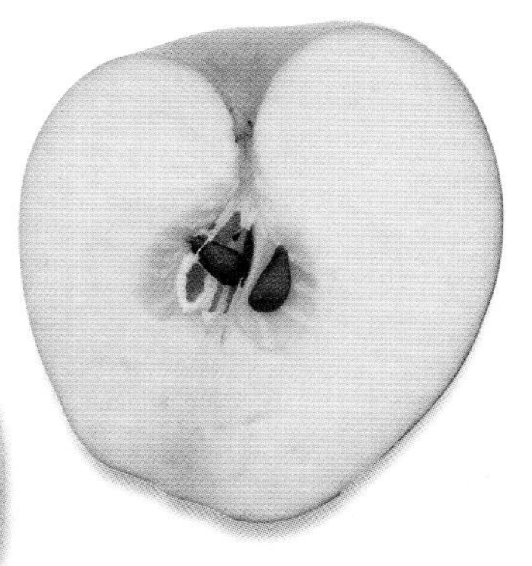

Type: Dessert.
Origin: Central Experimental Farm, Ottawa, Canada, 1928.
Parentage: Unknown.
Flowering: Early.

Note: Trees are very vigorous.

Right: Fruits of this variety have a distinctive shape and colouring.

Foster's Seedling

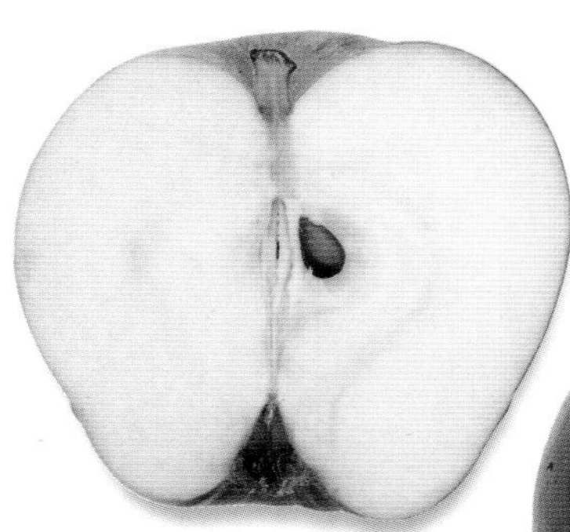

The slightly flattened fruits, around 6cm (2¼in) across, are bright green with red flushing and flecks. The white flesh is tender with an acid and vinous flavour.

Type: Culinary.
Origin: Maidstone, Kent, UK, *c.*1893.
Parentage: Unknown.
Flowering: Late.

Note: Fruits cook to a purée with a very sharp flavour. This variety is self-sterile so a compatible pollinator is required.

Fraise de Buhler

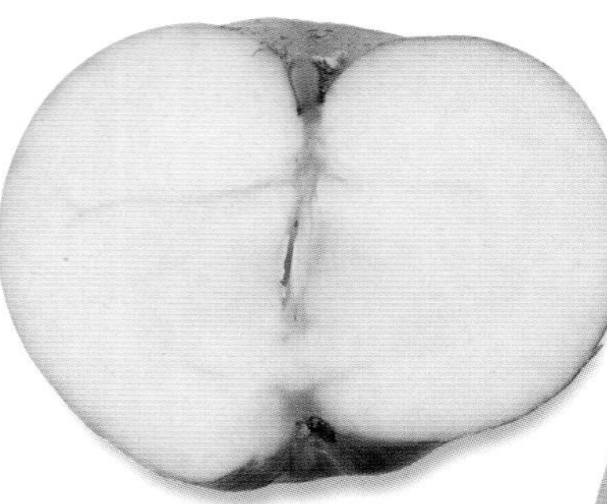

The very irregular fruits, around 7cm (2¾in) across, are yellow-green with a red flush and striping. The whitish flesh is firm and fine with a subacid flavour.

Type: Dessert.
Origin: ?Buhl, nr Baden, Germany, before 1947.
Parentage: Unknown.
Flowering: Early.

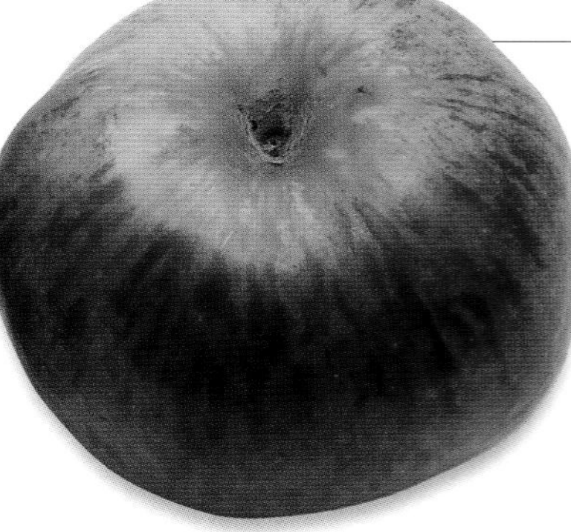

Note: The name of this variety suggests the fruits may have a strawberry-like flavour. Trees are very vigorous.

Franc-Bon-Pommier

Type: Dessert.
Origin: France, 1950.
Parentage: Unknown.
Flowering: Late.
Other name: Franc-Bon-Pommier (Moselle).

The somewhat flattened, slightly irregular fruits, around 6cm (2¼in) across, are green with a strong red flush and some striping. The white flesh is firm with a slightly sweet flavour.

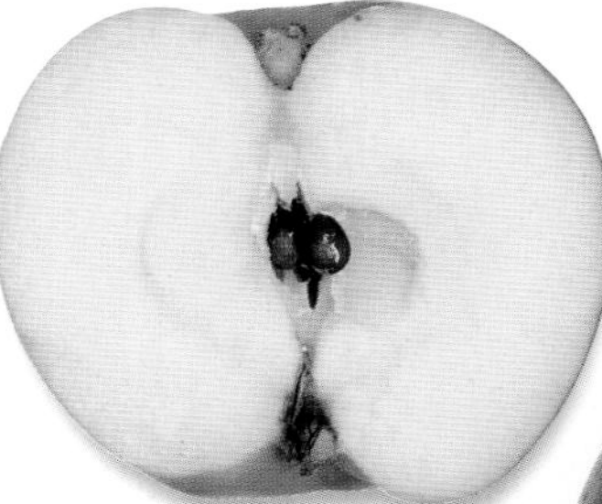

Note: The blossom of this variety, sometimes used for cider making, is susceptible to fireblight.

Left: The Franc-Bon-Pommier originates from the north of France.

France Deliquet

Type: Dessert.
Origin: Angers, France, 1950.
Parentage: Unknown.
Flowering: Late.

The somewhat knobbly and irregular, flattened fruits, around 7cm (2¾in) across, are bright yellow-green with a red flush and some russeting. The creamy white flesh is firm, fine and crisp with a slightly sweet, slightly subacid flavour.

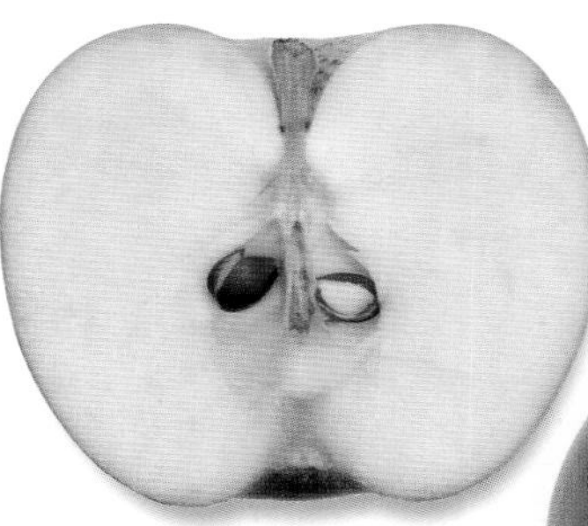

Note: Trees are moderately vigorous.

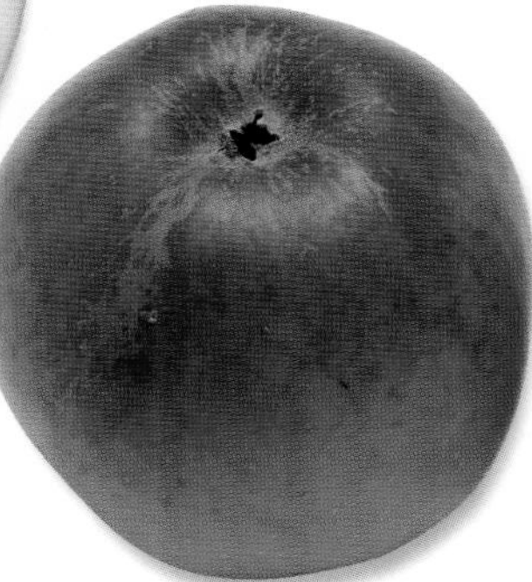

Above: Fruits are borne on spurs in clusters.

Francis

Type: Dessert.
Origin: Essex, UK, early 20th century.
Parentage: Cox's Orange Pippin (female) x Unknown.
Flowering: Mid-season.

The rounded to slightly conical fruits, around 6cm (2¼in) across, are bright yellow-green with a pinkish red flush and some striping and dotting. The creamy white flesh is firm, fine and crisp with a very sweet and aromatic flavour.

Note: Trees are moderately vigorous.

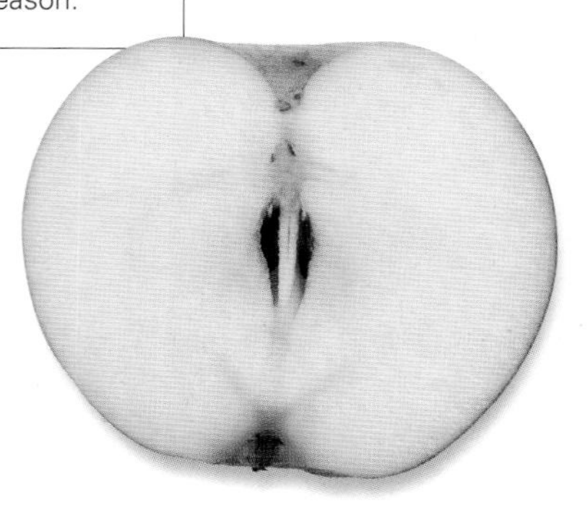

Freiherr von Berlepsch

The slightly flattened, somewhat irregular fruits, around 6cm (2¼in) across, are yellow-green with a pinkish red flush and some russeting. The creamy white flesh is crisp and very juicy with a subacid flavour.

Note: Trees are very vigorous, doing best in a sheltered position. They can be prone to canker and fungal diseases. Fruits can be stored for up to six months.

Type: Dessert.
Origin: Grevenbroich, Rheinland, Germany, *c.*1880.
Parentage: Ananas Reinette (female) x Ribston Pippin (male).
Flowering: Mid-season.
Other names: Freiherr de beri, Baron de Berlepsch, Berlepsch, Berlepschs Goldrenette, Freiherr von Berlepsch Gold-Reinette, Goldrenette Freiherr von Berlepsch, Reinette dorée de Berlepsch, Reinette Freiherr von Berlepsch, Renet Berlepsch and Renet zolotoi Berlepsha.

Frémy

The rounded fruits, around 6cm (2¼in) across, are clear green, generously splashed and marked with russeting over the entire surface. The yellowish flesh is firm and fine with a sweet, subacid and aromatic flavour.

Note: The flavour is similar to that of a Reinette apple. Trees are very vigorous.

Type: Dessert.
Origin: Chère, France, *c.*1830–40.
Parentage: Unknown.
Flowering: Unknown.
Other names: de Fremy and Gelineau.

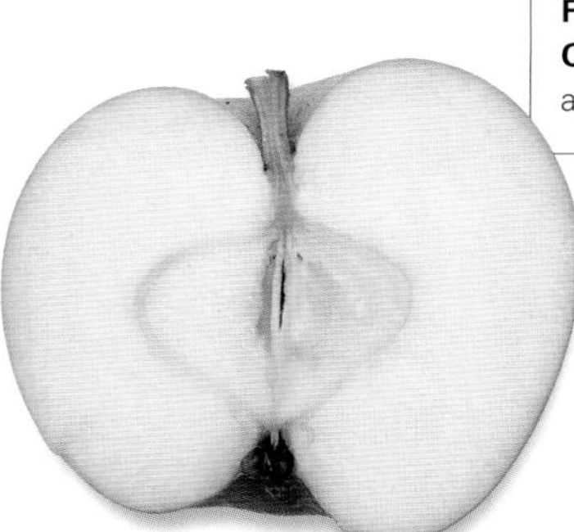

French Crab

The rounded fruits, around 6cm (2¼in) across, are bright green with some spotting and russeting and sometimes a dull red flush. The white flesh, distinctly tinged green, is very firm, coarse-textured and a little juicy and acid.

Note: Trees are very hardy and bear freely. The fruits cook well with a strong aroma. They can be stored for up to six months or even longer.

Left: *French crab is a cooking apple of fine quality.*

Type: Culinary.
Origin: Thought to be France, brought to England late 1700s.
Parentage: Unknown.
Flowering: Mid-season.
Other names: Amiens Long Keeper, Bobin, Claremont Pippin, Easter Pippin, Green Beefing, Grüner Oster, Iron King, Ironside, Ironstone, Ironstone Pippin, John Apple, Robin, Somerset Stone Pippin, Three Years Old, Tunbridge Pippin, Two Years Apple, Winter Greening, Winter Queening, Yorkshire Robin and Young's Long Keeper.

Fuji

Type: Dessert.
Origin: Horticultural Research Station, Nakahara, Japan, 1939 (named 1962).
Parentage: Ralls Janet (female) x Delicious (male).
Flowering: Mid-season.
Other name: Fuji INRA Type 4 – Nagafu (No.2) INFEL 6671.

The rounded fruits, around 6cm (2¼in) across, are yellow-green with a pink-red flush and flecking. The dull white flesh is crisp and juicy with a slightly subacid flavour.

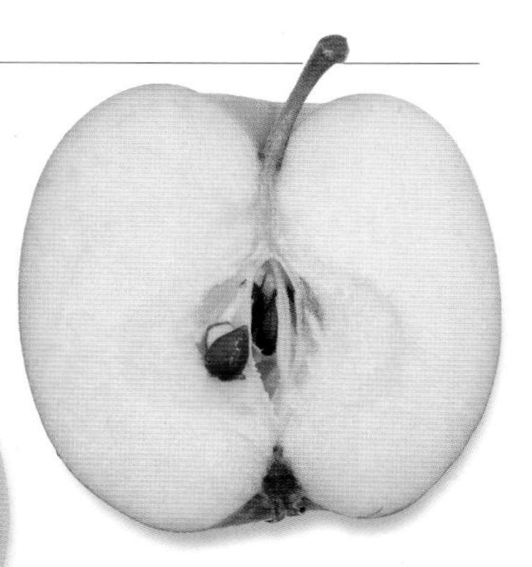

Note: This popular commercial variety does best in areas with warm summers. Scab and fireblight may be a problem in some areas. Fruits store well.

Fukunishiki

The rounded, slightly irregular fruits, around 6cm (2¼in) across, are yellow-green with a pinkish-red marbled flush. The creamy-white flesh is firm and crisp with a sweet flavour that is somewhat reminiscent of pear drop sweets.

Type: Dessert.
Origin: Aomori Apple Experiment Station, Japan, 1933 (named 1949).
Parentage: Ralls Janet (female) x Delicious (male).
Flowering: Mid-season.

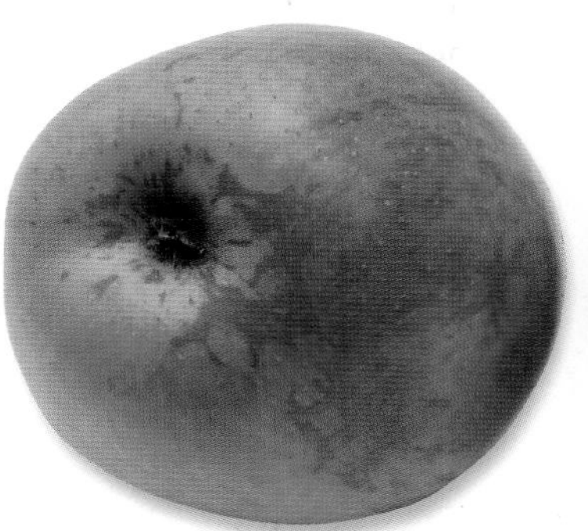

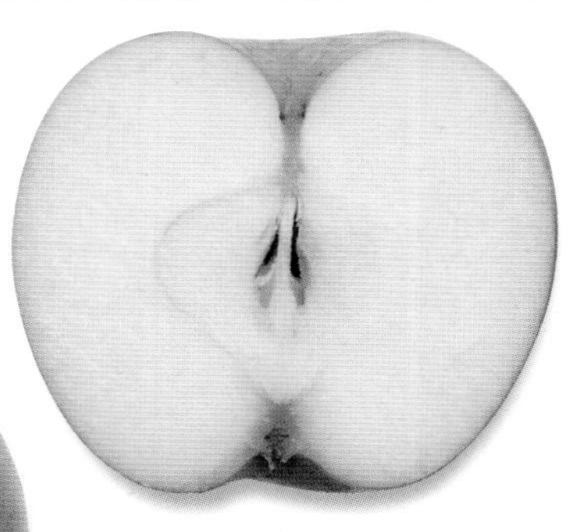

Note: Fruits need a long, warm summer to ripen fully.

Right: Fukunishiki is an attractive apple with a very distinctive taste when ripe.

Galloway Pippin

Type: Culinary.
Origin: Wigtown, Galloway, UK, 1871 (but thought to be much older).
Parentage: Unknown.
Flowering: Mid-season.
Other names: Croft en Reich, Croft St Andrews, Gallibro, Gallibro Pippin, Galloway, Galway's Pippin, Graft-en-Reich, Pepin Galloveiskii and Pepin Galloway.

The flattened, slightly uneven fruits, to around 7cm (2¾in) across, are bright yellow-green with some spotting. The creamy white flesh is firm, crisp and juicy with a sharp, subacid flavour.

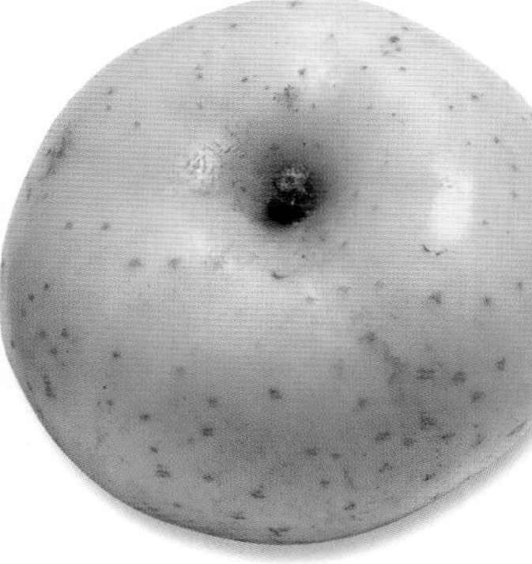

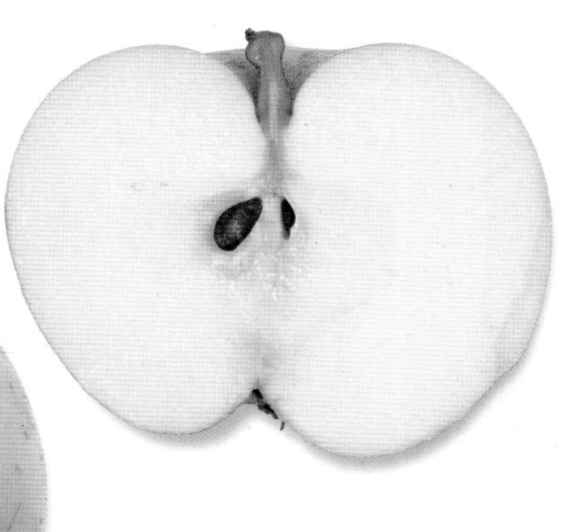

Note: The fruits of this old Scottish variety cook well but are best used early rather than storing.

Gambafina

Type: Dessert.
Origin: Carraglio, province of Cuneo, Italy, *c.*1900.
Parentage: Unknown.
Flowering: Mid-season.

The flattened, slightly irregular fruits, around 6cm (2¼in) across, are bright yellow with a strong dark red flush and some striping. The greenish white flesh is soft with a sweet subacid flavour.

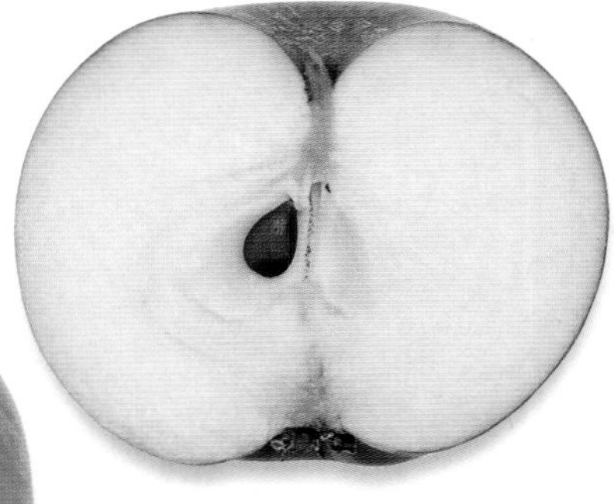

Note: This variety shows excellent disease resistance.

Gascoyne's Scarlet

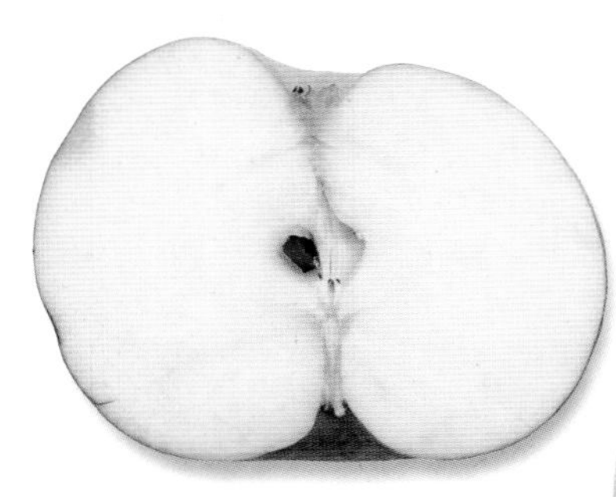

Note: This variety produces an attractive pink juice. Trees are very vigorous.

The very irregular fruits, around 6cm (2¼in) across, are light yellowish green with a pinkish red flush. The whitish flesh is firm, fine-textured and slightly juicy and sweet, but with very little flavour.

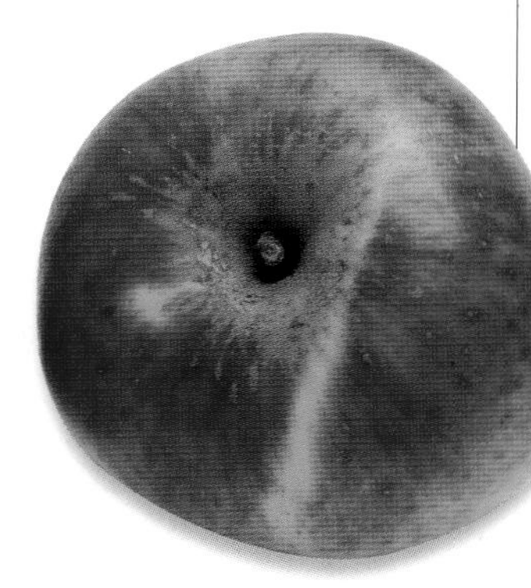

Type: Dessert and culinary.
Origin: Maidstone, Kent, UK, 1871.
Parentage: Unknown.
Flowering: Mid-season.
Other names: Cramoisie de Gascoigne, Friedrich August von Sachesen, Gascoigne's Scarlet, Gascoigne's Seedling, Malinovoe Gaskonskoe, Rhum von England and Schöner von Rusdorf.

George Carpenter

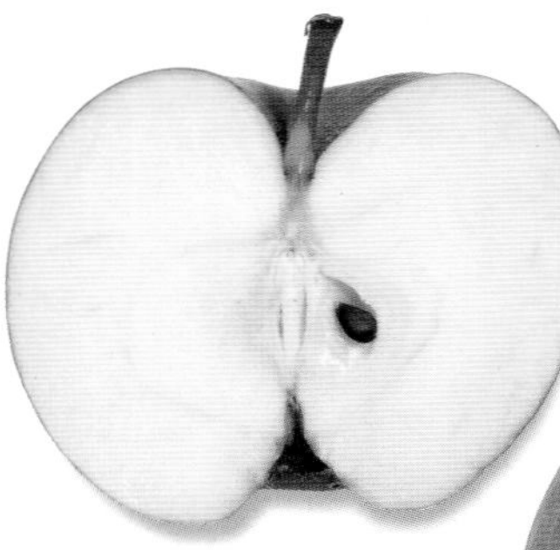

Note: The flavour is similar to Blenheim Orange. Trees are very vigorous.

The attractive, rounded fruits, around 8cm (3in) across, are yellow-green with a strong red flush. The white flesh is firm, fine-textured and juicy with a sweet, aromatic flavour.

Type: Dessert.
Origin: West Hall Gardens, Byfleet, Surrey, UK, 1902.
Parentage: Blenheim Orange (female) x King of the Pippins (male).
Flowering: Mid-season.

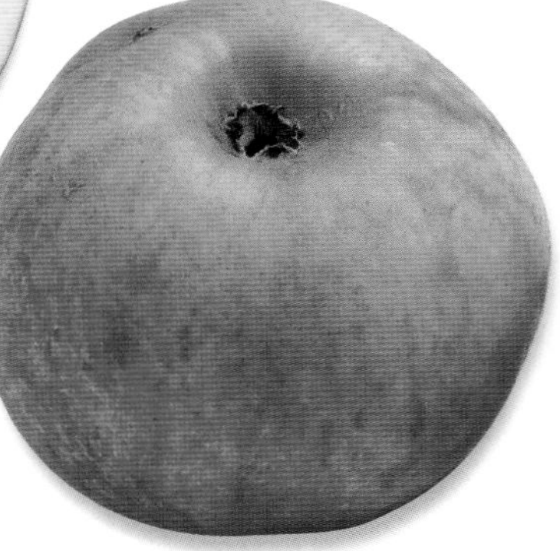

Gewürzluiken

Type: Dessert.
Origin: Germany, 1951.
Parentage: Unknown.
Flowering: Mid-season.

The rounded fruits are yellow-green with a strong red flush and striping. The whitish flesh is firm with a sweet subacid flavour.

Note: Flowers show some frost resistance. Fruits store well.

Giambun

Type: Dessert.
Origin: Italy, 1958.
Parentage: Unknown.
Flowering: Late.

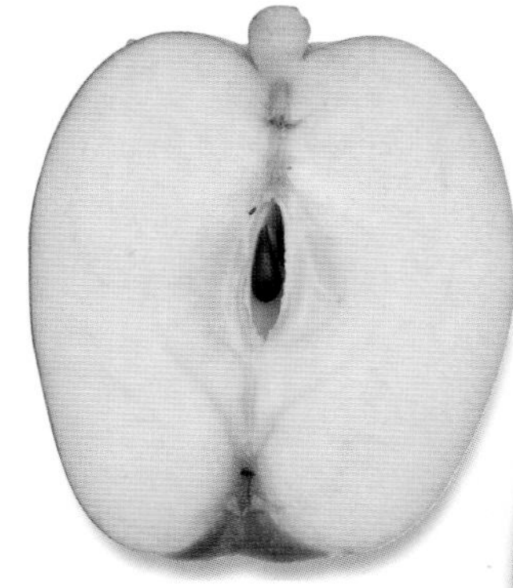

Note: Trees are moderately vigorous.

Left: The shape of the ripe fruits is very distinctive.

The long, conical fruits, around 5cm (2in) across, are bright yellow-green with a dark red flush. The creamy white flesh is firm and coarse with a subacid flavour.

Gian André

Type: Dessert.
Origin: Italy, 1958.
Parentage: Unknown.
Flowering: Late.

The conical, irregular fruits, around 7cm (2¾in) across, are bright yellow-green with a red flush. The creamy white flesh is firm, fine and tender with an insipid flavour.

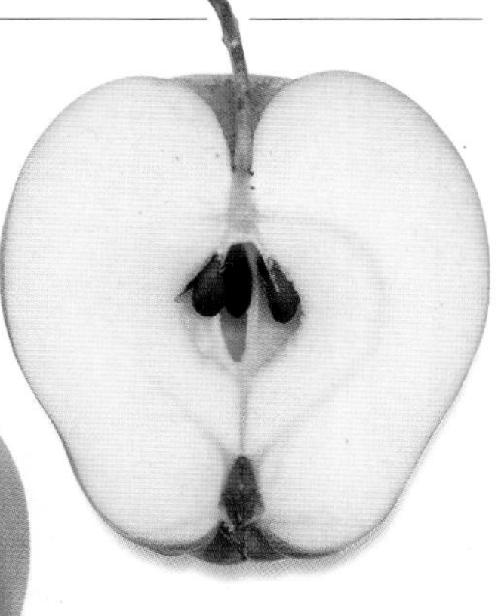

Above: Fruits of Gian André are slightly ridged and bumpy.

Note: Late flowering makes this a suitable variety for growing in cold districts. Trees are weak-growing.

Glengyle Red

The rounded fruits, around 7cm (2¾in) across, are bright yellow with a red flush. The creamy white flesh is coarse-textured and juicy but with little flavour.

Type: Dessert and culinary.
Origin: Balannah, South Australia, 1914.
Parentage: A more highly coloured sport of Rome Beauty.
Flowering: Late.

Note: The skins of this variety tend to be tough. Trees are moderately vigorous.

Gloria Mundi

The irregular, sometimes oblong, heavily ribbed fruits, often more than 8cm (3in) across, are bright green. The soft, creamy white flesh is coarse-textured and dry with a subacid taste.

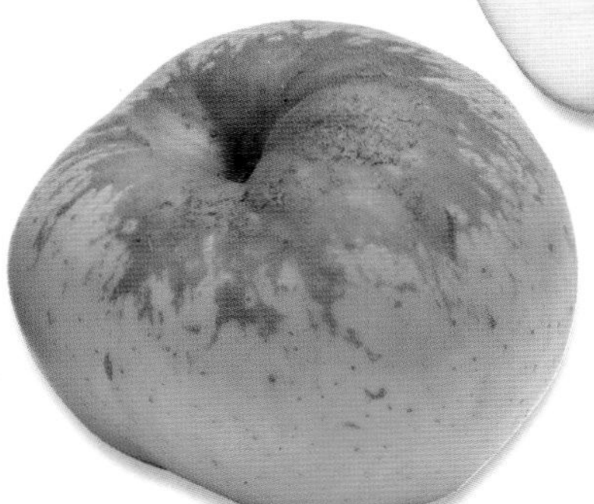

Note: The precise origins of this variety are in doubt owing to conflicting historical records.

Type: Culinary.
Origin: Germany or USA, 1804.
Parentage: Unknown.
Flowering: Mid-season.
Other names: American Mammoth, Baltimore, Belle Dubois, Belle Josephine, Copp's Mammoth, Glazenwood, Grosse de St Clement, Herrenapfel, Josephine, Kinderhook Pippin, Mammoth, Melon, Mississippi, Monstrous Pippin, Mountain Flora, Ox Apple, Pfundapfel, Pound, Ruhm der Welt, Slava Mira, Spanish Pippin, Titus Pippin and Vandyne Apple.

Golden Delicious

The slightly conical, sometimes irregular fruits, to 7cm (2¾in) across, are yellow. The creamy white flesh is crisp and juicy with a sweet, aromatic flavour.

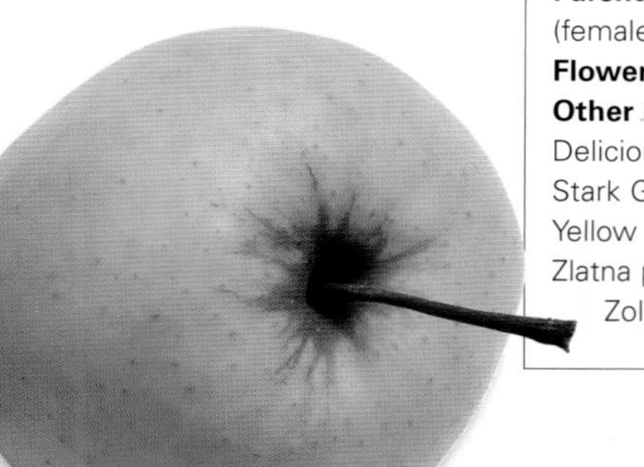

Type: Dessert.
Origin: Clay County, West VA, USA, 1905.
Parentage: ?Grimes Golden (female) x Unknown.
Flowering: Mid-season.
Other names: Arany Delicious, Delicios auriu, Stark Golden Delicious, Yellow Delicious, Zlatna prevazhodna and Zolotoe prevoshodnoe.

Note: This self-fertile variety does best in warm areas, where it is easy to grow. In some areas, scab, fireblight and rust may be a problem. Trees can be prone to biennial bearing.

Golden Knob

Type: Dessert.
Origin: Enmore Castle, Somerset, UK, late 1700s.
Parentage: Unknown.
Flowering: Mid-season.
Other names: Golden Nobb, Kentish Golden Knob, Old Lady and Old Maid.

The irregular fruits, around 5cm (2in) across, are bright yellow with heavy russeting. The flesh is firm with an intense, nutty, sweet-sharp flavour.

Note: Trees are vigorous and hardy and bear freely.

Golden Russet

Type: Culinary and dessert.
Origin: New York, USA, 1845.
Parentage: Seedling of English Russet.
Flowering: Early.

Note: Fruits can be stored for around three months. This variety is excellent for cider making. Trees are scab resistant.

The rounded fruits, of medium size, are greyish green to bronze with a coppery orange flush and russeting. The creamy white flesh is fine-grained, crisp and juicy with a sweet flavour.

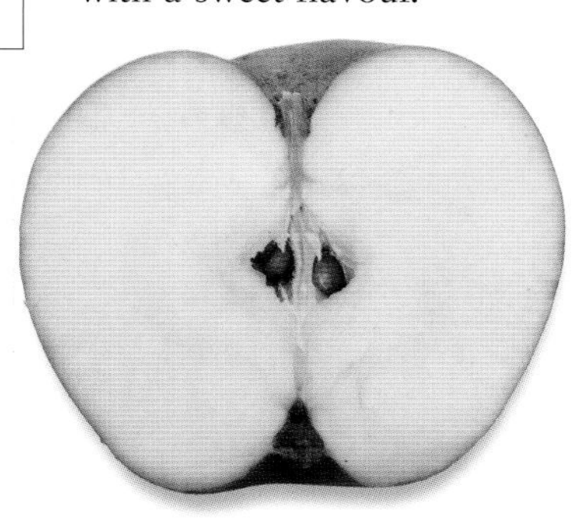

***Right**: Golden Russet is an excellent eating apple of fine quality.*

Goldjon

Type: Dessert.
Origin: Turin University, Italy, date not recorded.
Parentage: Golden Delicious (female) x Jonathan (male).
Flowering: Mid-season.

The slightly conical, somewhat irregular fruits, around 7cm (2¾in) across, are bright yellow-green with a pinkish red flush. The creamy white flesh is crisp and juicy with a sweet flavour.

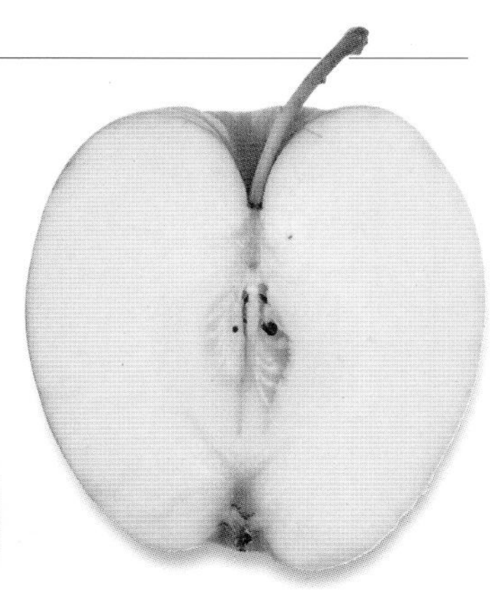

Note: This variety is suitable for juicing.

***Far left**: The colouring of the ripe fruits is attractive.*

Granny Giffard

The rounded to conical fruits, around 6cm (2¼in) across, are light green with a pinkish red flush and some striping and spotting. The yellowish white flesh is fine and tender with a subacid flavour.

Type: Culinary and dessert.
Origin: Minster, near Margate, Kent, UK, exhibited 1858.
Parentage: Unknown.
Flowering: Late.
Other name: Granny Gifford.

Note: Fruits can be stored for three to four months. Trees are moderately vigorous.

Granny Smith

The rounded fruits, around 6cm (2¼in) across, are an even bright green with lighter flecking. The creamy white flesh is firm, rather coarse-textured and juicy with a refreshing, subacid flavour.

Note: This variety, which does best in warm areas, was raised by Mrs Thomas Smith. Born in Peasmarsh, Sussex, UK, in 1800, she emigrated to Australia in 1838. In some areas, powdery mildew and fireblight may cause problems. The fruits store well.

Type: Dessert.
Origin: Ryde, NSW, Australia, before 1868.
Parentage: ?French Crab (female) x Unknown.
Flowering: Early.

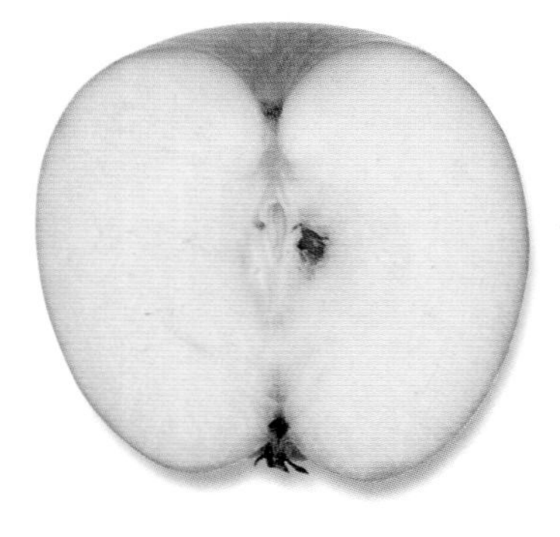

Grantonian

The rounded fruits, more than 8cm (3in) across, are bright green. The creamy white flesh is coarse, mealy and soft with a slightly subacid flavour.

Note: This variety is self-sterile so a compatible pollination partner is required for successful fruiting. The fruits store well.

Type: Culinary.
Origin: Nottingham, UK, 1883.
Parentage: Unknown.
Flowering: Mid-season.

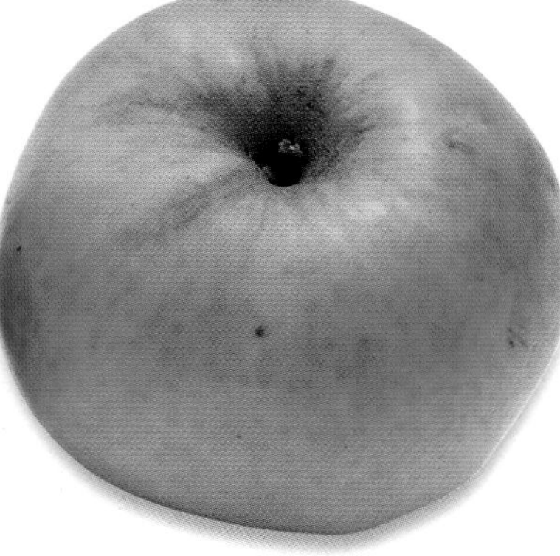

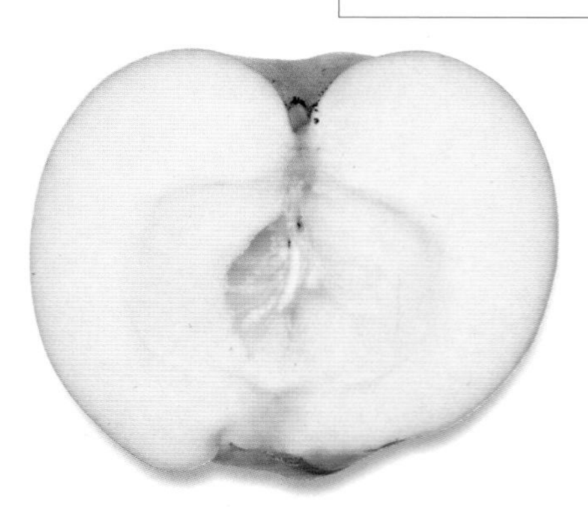

Gravenstein

Type: Culinary and dessert.
Origin: Schleswig-Holstein, Italy or Southern Tyrol; arrived in Denmark *c.*1669.
Parentage: Unknown.
Flowering: Early.
Other names: A. Grafenstein, Blumencalvill, Calville Gravenstein, de Comte, Diels Sommerkönig, Early Congress, Ernteapfel, Gelber Gravensteiner, Graasten, Gravstynke, Ohio Nonpareil, Paradiesapfel, Petergaard, Prinzessinapfel, Rippapfel, Romarin de Botzen, Sabine, Stroemling, The Gravenstein Apple and Tom Harryman.

The thin-skinned, somewhat uneven fruits, up to 8cm (3in) across, are yellow-green with orange-red mottling and streaking. The yellowish white flesh is crisp, rather coarse-textured and juicy with a pleasant mixture of sweetness and acidity and a distinctive flavour.

Note: This variety is a triploid. It was declared the national apple of Denmark in 2005. Trees are vigorous, developing a large crown, and are prone to biennial bearing. Susceptible to leaf spots, fungal disease and bitter pit but resistant to canker. Fruits can be stored for around two to three months.

***Left**: Gravenstein is a historic variety that has remained in cultivation.*

Green Purnell

Type: Dessert.
Origin: Worcestershire, UK. Recorded 1945 but believed to be much older.
Parentage: Unknown.
Flowering: Mid-season.

The irregular, somewhat flattened fruits, around 7cm (2¾in) across, are bright green with a red flush. The creamy white flesh is fine-textured with a slightly sweet, subacid flavour.

Note: Though rare in cultivation nowadays, this variety has been used in the restoration of old orchards in the British Midlands. Trees are moderately vigorous.

***Far left**: Fruits can show some spotting and russeting.*

Greensleeves

Type: Dessert.
Origin: East Malling Research Station, Maidstone, Kent, UK, 1966.
Parentage: James Grieve (female) x Golden Delicious (male).
Flowering: Very early.

The rounded, occasionally oblong to conical fruits, around 7cm (2¾in) across, are bright yellow-green with some russeting. The creamy white flesh, sometimes tinged pink-orange, is crisp and juicy with a mild, refreshing taste.

Note: This is a prolific variety. Trees are partially tip-bearing and bear fruit when young. Early flowering makes it unsuitable for use in very cold areas.

***Right**: Greensleeves is a good garden apple with a pleasant flavour.*

Grimes Golden

Type: Dessert.
Origin: Brook County, West VA, USA, 1804.
Parentage: Unknown.
Flowering: Early.
Other names: Dorée de Grimes, Grimes, Grimes Goldapfel, Grimes Golden Pippin, Grimes Yellow Pippin and Zolotoe Graima.

The rounded fruits, around 7cm (2¾in) across, are bright greenish yellow. The creamy white flesh is crisp, juicy and fine-textured with a moderately sweet and flavour.

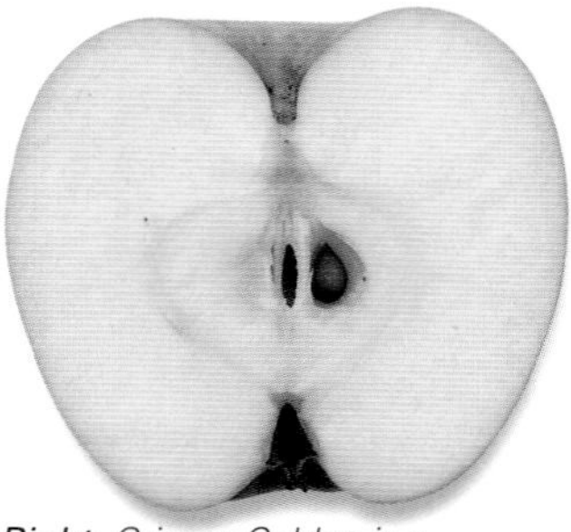

Right: Grimes Golden is a probable parent of the popular Golden Delicious variety.

Note: Young trees bear reliably and show some resistance to fireblight and apple cedar rust. The fruits are suitable for cider making.

Groninger Kroon

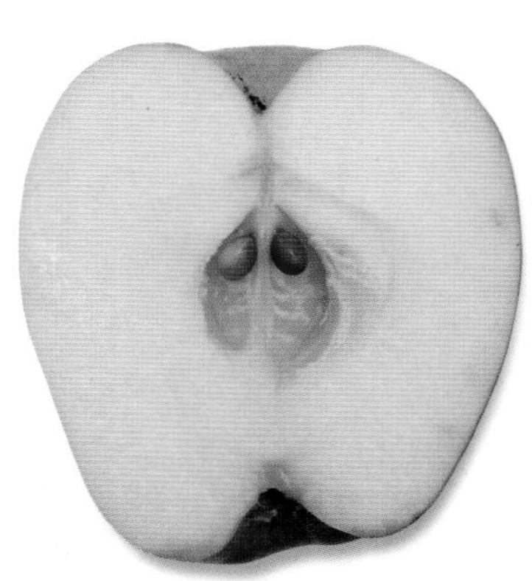

The conical, sometimes almost oblong fruits, around 6cm (2¼in) across, are yellowish green with a pinkish red flush. The white flesh is very fine and firm with a slightly sweet flavour.

Type: Dessert.
Origin: Netherlands, before 1944.
Parentage: Unknown.
Flowering: Mid-season.
Other name: Groningen Kroon.

Note: This variety is resistant to apple scab. Trees are moderately vigorous.

Gronsvelder Klumpke

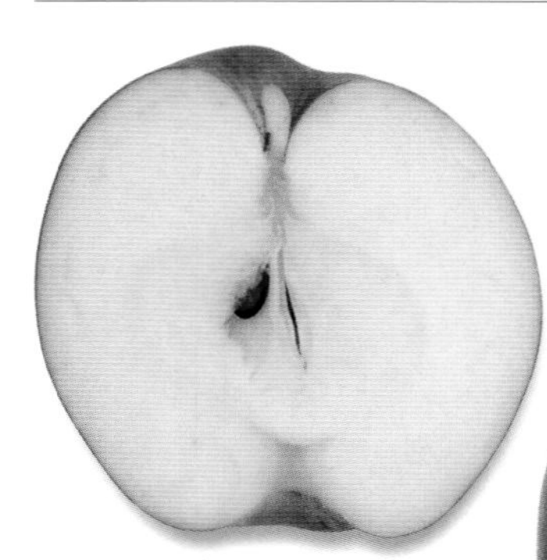

The somewhat irregular, slightly oblong fruits, around 6cm (2¼in) across, are bright green with a strong dark red flush. The creamy white flesh is firm and fine-textured with a subacid flavour.

Type: Dessert.
Origin: Netherlands and Belgium, 1948.
Parentage: Unknown.
Flowering: Late.
Other names: Rood Klumpke and Sabot de Gronsveld.

Note: This variety is a sport of Eijsdener Klumpke. Trees are very vigorous.

Gros-Api

Type: Dessert.
Origin: Brittany, France, recorded 1628.
Parentage: Unknown.
Flowering: Mid-season.
Other names: Api Blanc, Api Bolshoe, Api Double, Api Grande, Api Grosse, Api rose, Dieu, Double Api, Double Rose, Drap d'Or (Villeneuve d'Agen), Gros Api, Grosser Api, Large Lady Apple, Passe Rose, Poma Rosa, Rose de l'Angenais, Rose de Provence, Rose Double Api, Rosenapfel, Rosenapi, Rubenapfel, Rubin and Vermillon Rubis.

The flattened, irregular fruits, up to 7cm (2¾in) or more across, are bright yellowish green with a pinkish red flush and some russeting. The white flesh is firm and fine-textured with a sweet, subacid and perfumed flavour.

Note: Fruits are wind-resistant. Mature trees crop freely.

Gros-Locard

Type: Culinary and dessert.
Origin: France, before 1849.
Parentage: Unknown.
Flowering: Late.
Other names: de Locard, Gro-Lokar, Gros-Locar, Locard Bicolore, Locard Groseille and Pomme de Locard.

The irregular fruits, around 7cm (2¾in) across, are bright yellow-green with some russeting. The creamy white flesh is crisp with a sweet and slightly acid flavour.

Note: Trees are vigorous. The fruits are suitable for juicing.

Grosse Mignonnette d'Herbassy

Type: Dessert.
Origin: France, first described 1934.
Parentage: Unknown.
Flowering: Mid-season.
Other names: Cabassou, Demoiselle and Mignonnette d'Herbassy.

The flattened fruits, around 7cm (2¾in) across, are bright yellow-green with a red flush. The flesh is fairly firm with a subacid and slightly sweet flavour.

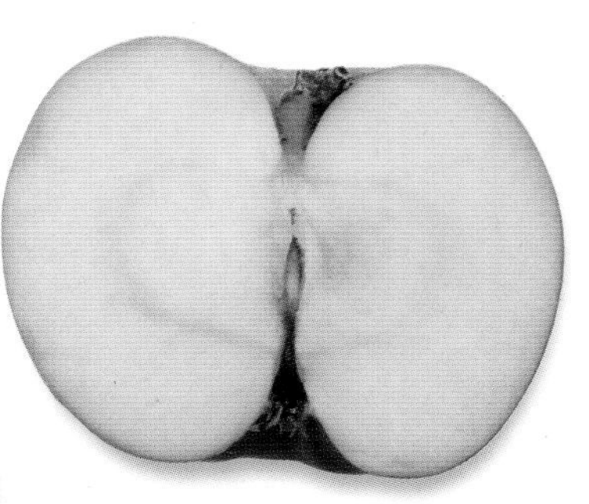

Note: Fruits can be stored for several months.

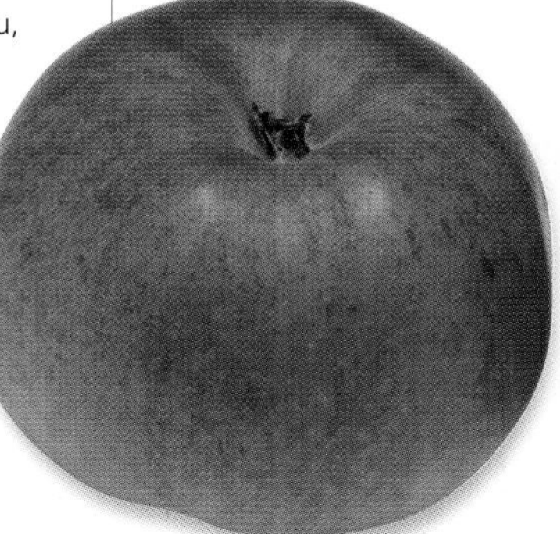

Grvena Lepogvetka

The conical, almost pear-shaped fruits, around 6cm (2¼in) across, are bright yellow with a dark red flush and some striping. The creamy white flesh has a sweet, subacid flavour.

Type: Dessert.
Origin: Former Yugoslavia, 1975.
Parentage: Unknown.
Flowering: Mid-season.

Note: Trees are moderately vigorous.

Gustavs Dauerapfel

The somewhat oblong fruits, around 6cm (2¼in) across, are bright yellow-green with a strong red flush. The creamy white flesh is firm and fine-textured with a sweet, slightly subacid and perfumed flavour.

Note: Fruits store for up to five months. Trees are moderately vigorous.

Type: Dessert.
Origin: Wadenswil, Switzerland, first described 1899.
Parentage: Unknown.
Flowering: Early.
Other names: Gustav Dauerapfel, Gustav Durabil and Gustavovo trvanlive.

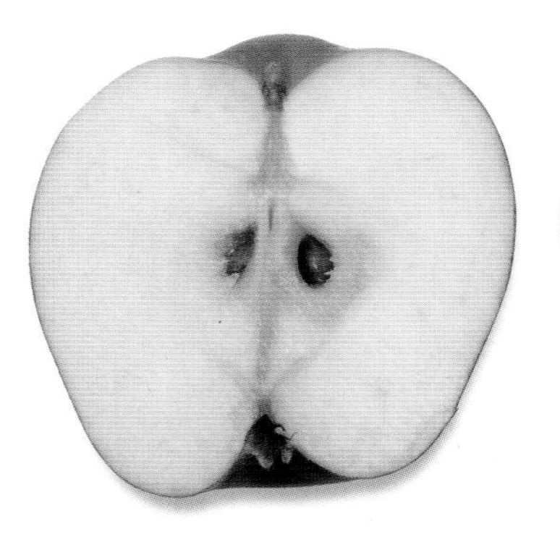

Gyógyi Piros

The rounded fruits, around 7cm (2¾in) across, are bright greenish yellow with a strong dark red flush and some flecking. The creamy white flesh is fairly firm and fine-textured with a subacid and slightly sweet flavour.

Type: Dessert.
Origin: Romania or Hungary, first recorded 1860.
Parentage: Unknown.
Flowering: Early.
Other names: Gyogyer roter, Roter Gyogger and Royii de Geoagiu.

Note: Trees are moderately vigorous.

Hamvas Alma

The somewhat irregular fruits, to 8cm (3in) across, are bright greenish yellow with a red flush and some flecking. The creamy white flesh is fairly fine and soft with a sweetish subacid flavour.

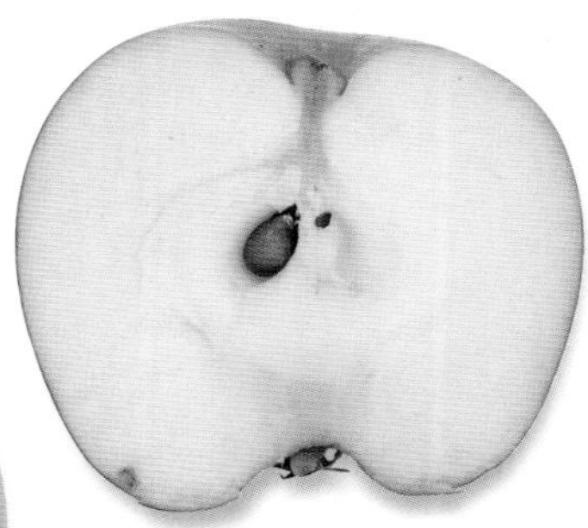

Type: Dessert.
Origin: Hungary, 1948.
Parentage: Unknown.
Flowering: Mid-season.

Note: Trees are moderately vigorous.

Right: Trees bear soft-fleshed fruit with plenty of fruity acidity.

Haralson

Type: Dessert.
Origin: Excelsior, MN, USA, selected 1913 and introduced 1923.
Parentage: Malinda (female) x Unknown.
Flowering: Mid-season.

The rounded fruits, around 7cm (2¾in) across, are bright yellow-green with a strong dark red flush. The creamy white flesh is crisp and juicy.

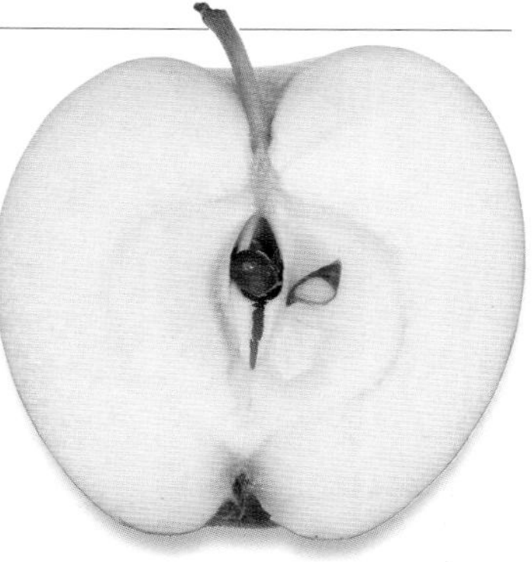

Note: The fruits can also be cooked and hold their shape well. Stored fruits retain their flavour well. Trees are vigorous with a tendency to biennial bearing.

Harry Pring

Type: Dessert.
Origin: Surrey, UK, 1911.
Parentage: Unknown.
Flowering: Mid-season.

The rounded fruits, around 6cm (2¼in) across, are yellow-green with a pinkish- to orange-red flush. The creamy white flesh has a savoury flavour.

Note: Trees are moderately vigorous. This variety is self-sterile so a compatible pollination partner is needed for good fruiting.

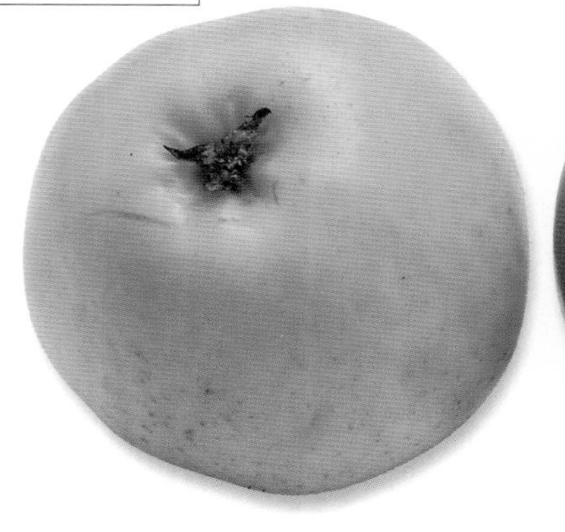

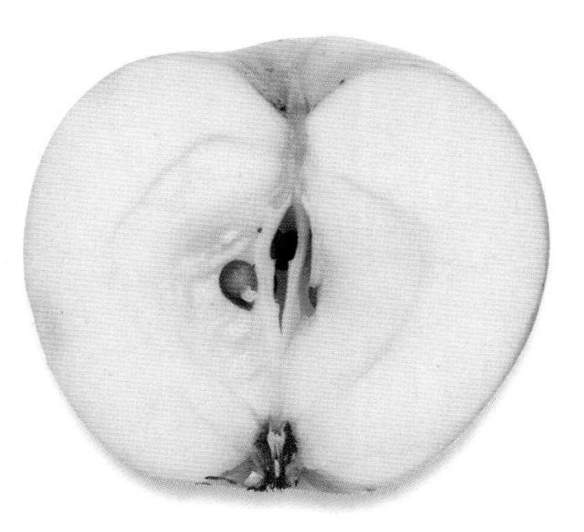

Harvey

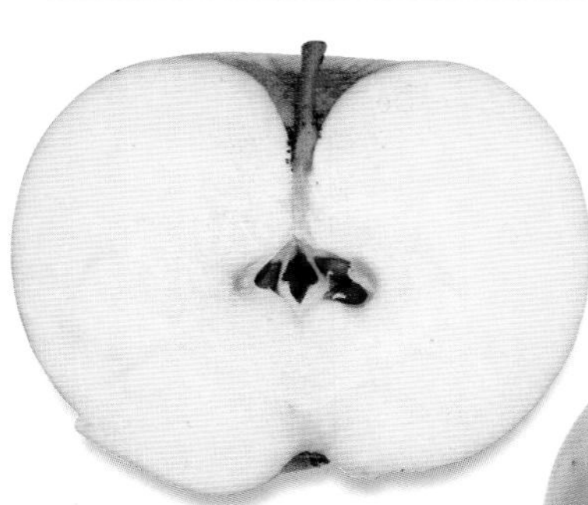

The irregular, slightly flattened fruits, around 7cm (2¾in) across, are bright yellow-green with some russeting. The creamy white flesh is firm, coarse-textured and very dry with a subacid and perfumed flavour.

Type: Culinary.
Origin: Norfolk, UK, first mentioned 1629 by English botanist Parkinson.
Parentage: Unknown.
Flowering: Mid-season.
Other names: Doctor Harvey, Doctor Harvey's Apple, Golden Warrior, Harvey Apple and The Doctor.

Note: This variety is of considerable historic interest. Trees are moderately vigorous.

Hawthornden

The flattened fruits, around 7cm (2¾in) across, are bright greenish yellow with a red flush. The creamy white flesh is firm and coarse with a subacid flavour.

Type: Culinary.
Origin: Scotland, UK, 1780.
Parentage: Unknown.
Flowering: Mid-season.
Other names: Apfel von Hawthornden, Epine blanche, Glogowka, Hagendornsapfel, Haley, Hawley, Hawthornden Old, Hawthornden Red, Hawthornden White, Hawthorndenske, Hlohovske, Lincolnshire Pippin, Lord Kingston, Shoreditch White, Weeler's Kernel, Weisser Hawthornden, Wheeler's Kernel, White Apple and White Hawthornden.

Note: Trees are generally healthy and vigorous and crop freely. Canker and woolly aphids may cause problems.

Hejocsabai Sarga

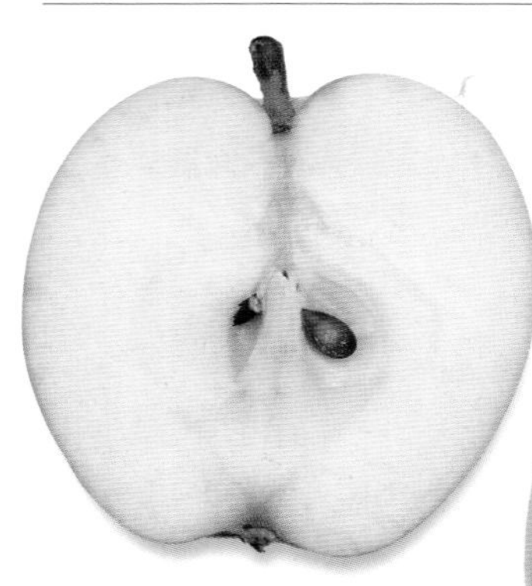

The rounded fruits, around 7cm (2¾in) across, are bright yellow-green. The very white flesh has a sweet, almost scented flavour.

Type: Dessert.
Origin: Hungary, 1948.
Parentage: Unknown.
Flowering: Mid-season.

Note: Hejocsabai is a district in Miskolc, an industrial city in north-eastern Hungary.

Herefordshire Beefing

The somewhat flattened, slightly irregular fruits, around 7cm (2¾in) across, are bright yellow-green with a strong orange to dark red flush and some russeting and spotting. The yellowish white flesh is firm and fine-textured with a moderately acid flavour.

Type: Culinary.
Origin: Herefordshire, UK, known in the late 1700s.
Parentage: Unknown.
Flowering: Late.
Other names: Hereford Beaufin, Hereford Beefing, Herefordshire and Herefordshire Beaufin.

Note: Fruits can be stored for around two to three months. They are very heavy for their size.

Left: Ripening fruits are dark and shiny.

High View Pippin

Type: Dessert.
Origin: Ernest Hill, Weybridge, Surrey, UK, 1911.
Parentage: Sturmer Pippin (female) x Cox's Orange Pippin (male).
Flowering: Mid-season.

The slightly conical fruits, around 7cm (2¾in) across, are dull yellow-green with a broken red flush. The creamy white flesh is firm, fine-textured and juicy with a sweet, pleasant, aromatic flavour.

Note: Fruits store well. Trees are moderately vigorous.

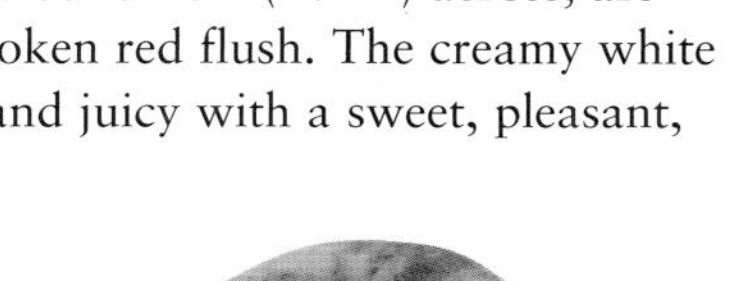

Histon Favourite

Type: Dessert and culinary.
Origin: Histon, Cambridgeshire, UK, first recorded 1883.
Parentage: Unknown.
Flowering: Early.
Other names: Chiver's Seedling, Chivers' Seedling and Histon Favorite.

The somewhat flattened, sometimes irregular fruits, around 7cm (2¾in) across, are yellow-green with a pinkish red flush. The creamy white flesh is rather soft, fine textured and juicy flesh with a faint flavour that is only a little sweet.

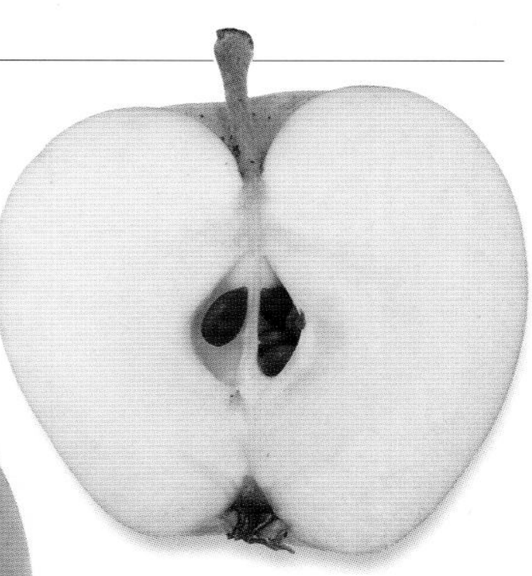

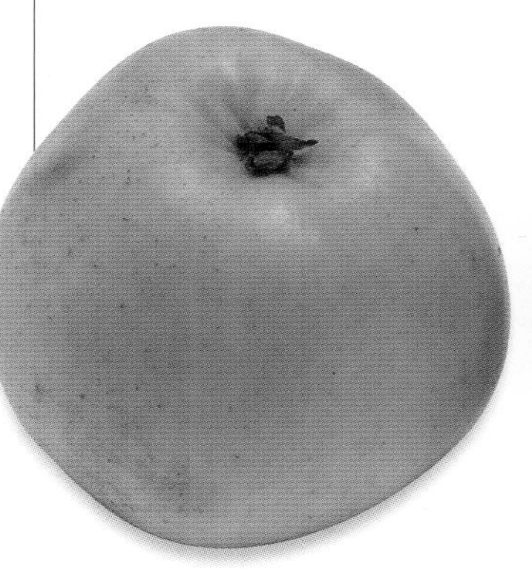

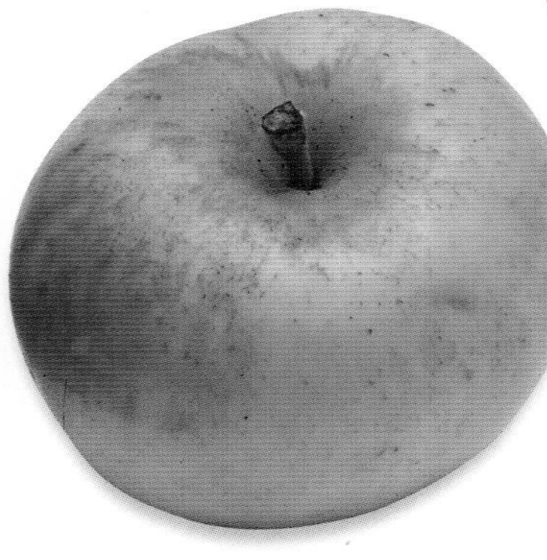

Note: Fruits can be stored for two to three months. Trees are moderately vigorous.

Hoary Morning

Type: Culinary.
Origin: ?Somerset, UK, first recorded 1819.
Parentage: Unknown.
Flowering: Mid-season.
Other names: Bachelor's Glory, Bedu Pteter Morgen Apfel, Blendon Seedling, Brouillard, Dainty, Downey, Downy, General Johnson, Harmat alma, Honeymoon, Mela pruinosa, Morgenduft, New Margil, Pruhaty ploskoun, Sam Rawlings, Utrennyaya rosa and Webster's Harvest Festival.

The flattened fruits, around 7cm (2¾in) across, are yellowish green with a red flush that often appears as striping. The whole surface is covered with a thick bloom, like frost. The creamy white flesh is firm, rather coarse-textured and dry with no flavour and little acidity.

Note: Fully ripe fruits are also edible raw. Trees are moderately vigorous.

Hog's Snout

The knobbly, irregular fruits, around 7cm (2¾in) across, are bright yellow-green with some russeting. The creamy white flesh is soft with a slightly acid flavour.

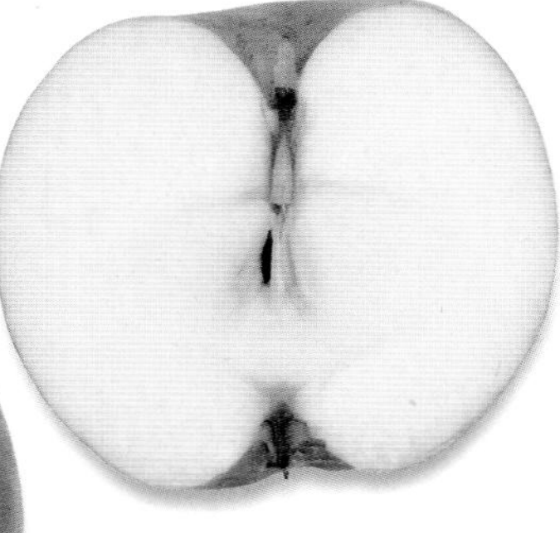

Type: Dessert.
Origin: UK, 1947.
Parentage: Unknown.
Flowering: Early.

Note: Trees are moderately vigorous.

Right: This variety is grown almost as much for the charm of its name as for the fruit itself.

Hohenzollern

Type: Dessert.
Origin: Probably Germany, 1947.
Parentage: Unknown.
Flowering: Mid-season.

The slightly flattened, somewhat irregular fruits, around 7cm (2¾in) across, are bright yellow-green with some russeting. The flesh is crisp and coarse with a slightly sweet flavour.

Note: Some sources indicate a French origin for this variety. Trees are very vigorous.

Far right: This apple commemorates a notable German aristocratic family.

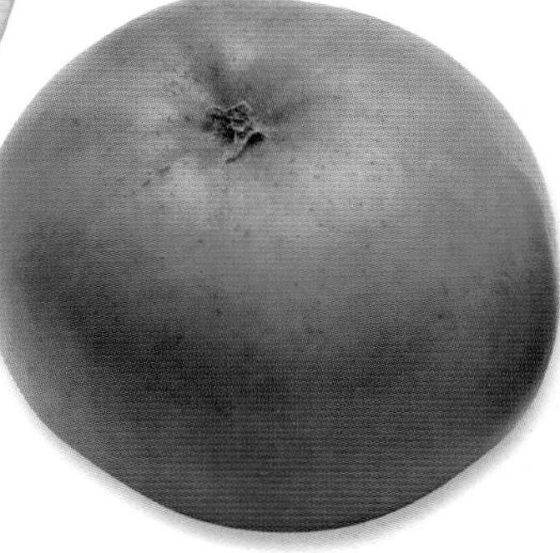

Holstein

Type: Dessert.
Origin: Eutin, Holstein, Germany, *c.*1918.
Parentage: Unknown, but possibly involving Cox's Orange Pippin.
Flowering: Early.
Other names: Holstein Cox, Holsteiner Cox, Holsteiner Gelber Cox and Vahldiks Cox Seedling No.III.

The somewhat oval but irregular fruits, to 8cm (3in) across, are yellow-green with a broken red flush. The creamy white flesh is firm, slightly coarse-textured and juicy with a little acidity and a sweet, richly aromatic flavour.

Note: This variety is a triploid so needs two pollinators. Fruits are suitable for juicing. Trees are vigorous and resistant to scab but the flowers are susceptible to frost damage.

Horei

Type: Dessert.
Origin: Aomori Apple Experiment Station, Japan, 1931, introduced 1949.
Parentage: Ralls Janet (female) x Golden Delicious (male).
Flowering: Mid-season.

The slightly conical fruits, around 7cm (2¾in) across, are bright yellow-green with an even red flush. The creamy white flesh is very firm with a fairly sweet flavour.

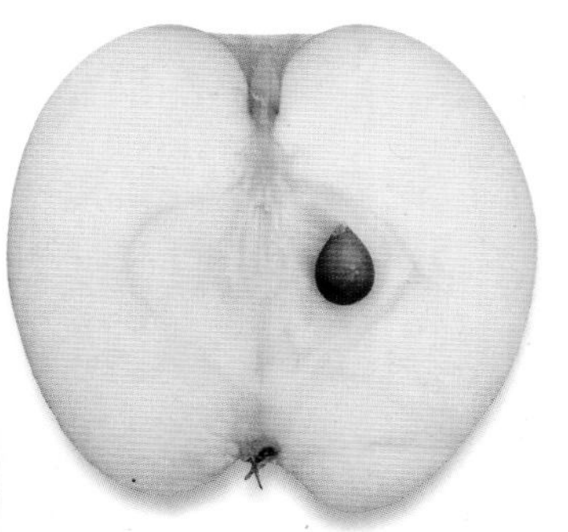

Note: The skins of the fruits are high in vitamins. Trees are weak growing.

Houblon

Type: Dessert.
Origin: Welford Park, Newbury, Berkshire, UK, first recorded 1901.
Parentage: Peasgood's Nonsuch (female) x Cox's Orange Pippin (male).
Flowering: Mid-season.

The rounded fruits, around 7cm (2¾in) across, are bright green with a red flush and striping. The creamy white flesh is firm, slightly coarse-textured and moderately juicy with a little acidity and a good aromatic, slightly aniseed taste.

Note: This variety is self-sterile so a suitable pollination partner is needed for good fruiting. Trees are moderately vigorous.

Howgate Wonder

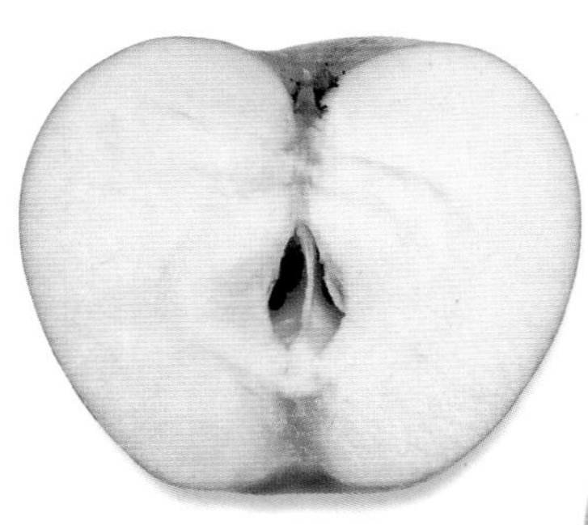

The rounded but sometimes uneven fruits, often much more than 7cm (2¾in) across, are bright greenish yellow with a red flush and striping. The creamy white flesh is firm, fine-textured and juicy, quite sweet when ripe and with a faint aromatic flavour.

Type: Culinary.
Origin: Bembridge, Isle of Wight, UK, 1915–16, introduced 1932.
Parentage: Blenheim Orange (female) x Newton Wonder (male).
Flowering: Mid-season.

Note: The fruits cook well. Fully ripe, they are also edible raw. This variety is partially self-fertile. Trees are vigorous and crop heavily.

Hubbardston Nonsuch

The rounded but sometimes uneven fruits, around 7cm (2¾in) across, are bright greenish yellow with a strong bright red flush. The creamy white flesh is firm and fine-textured with a sweet, subacid flavour.

Type: Dessert.
Origin: Hubbardston, MA, USA, first recorded 1832.
Parentage: Unknown.
Flowering: Mid-season.
Other names: American Blush, American Nonpareil, Farmer's Profit, Hubardston Pippin, Hubbardston Old Town Pippin, Hubbardston Pippin, John May, Monstreuse d'Amerique, Nonesuch, Nonpareille de Hubbardston, Nonsuch, Old Town Pippin, Orleans, Sans Pareille d'Hubbardston, Sondergleichen von Hubbardston and Van Vleet.

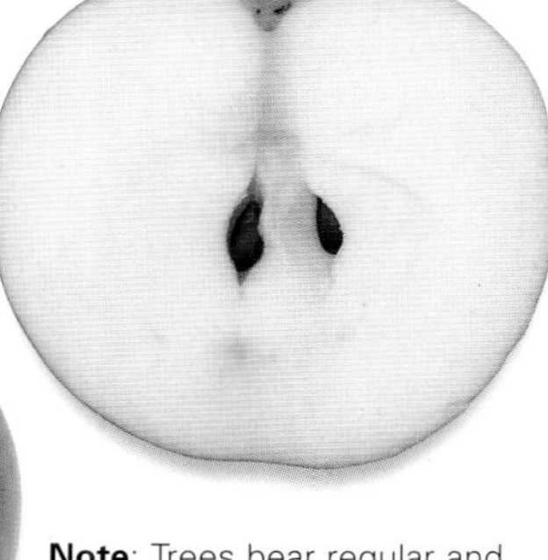

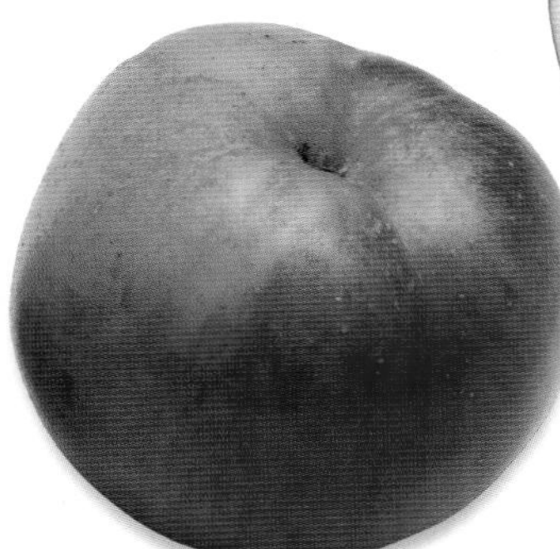

Note: Trees bear regular and heavy crops of evenly sized fruits. Trees are moderately vigorous.

Idared

The rounded fruits, around 7cm (2¾in) across, are bright green with a strong bright crimson-red flush and some indistinct streaking. The white, green- or pink-tinged flesh is firm, crisp and fine-textured with a sweet and pleasant vinous flavour.

Type: Dessert and culinary.
Origin: Idaho Agricultural Experiment Station, Moscow, USA, introduced 1942.
Parentage: Jonathan (female) x Wagener (male).
Flowering: Early.

Right: This sweet red apple is renowned for its keeping qualities.

Note: Trees tolerate heavy soil. They can be susceptible to fungal diseases. Young trees crop well. For the best colour, fruits need good exposure to sun. They store well over a long period.

Ildrod Pigeon

Type: Dessert.
Origin: Island of Fyn, Denmark, *c.*1840.
Parentage: Unknown.
Flowering: Early.
Other names: Eldrau Pigeon, Eldrod Pigeon, Eldrott Duvapple, Feuerroter Taubenapfel, Golubok ognenno-krasnyi, Ilrood Pigeon, Morke rod Pigeon, Pigeon rouge feu, Taubenapfel Feuerroter and Yldrod Pigeon.

The somewhat oblong fruits, around 7cm (2¾in) across, are bright greenish yellow with a red flush. The white flesh is firm, crisp and fine-textured with a sweet, subacid and slightly aromatic flavour.

Note: This variety is susceptible to powdery mildew. Some viruses can also cause problems.

Ingol

Type: Dessert.
Origin: Jork Fruit Research Station, Hamburg, Germany, 1954.
Parentage: Ingrid Marie (female) x Golden Delicious (male).
Flowering: Mid-season.

The flattened fruits, around 7cm (2¾in) across, are greenish yellow with a pinkish red flush and some striping. The creamy white flesh is soft but juicy with a rich flavour.

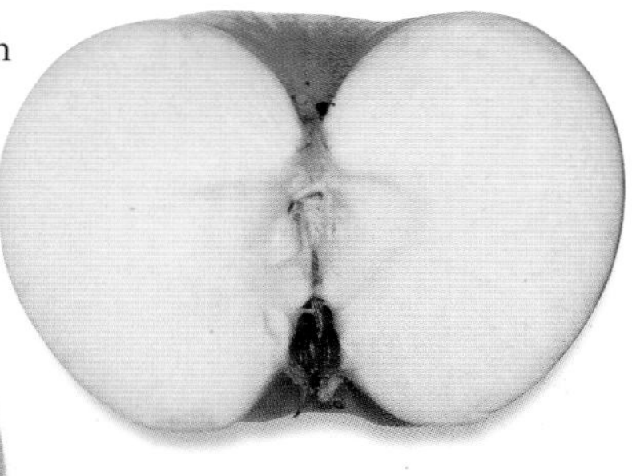

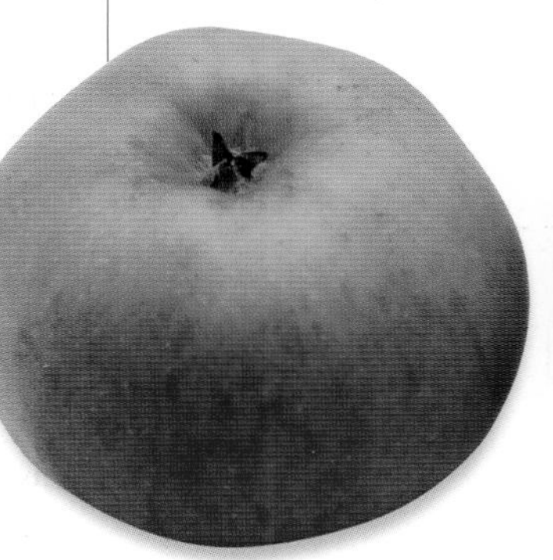

Note: This variety is resistant to bitter pit. Fruits store very well but lose flavour after four months. Trees are moderately vigorous.

Ingrid Marie

Type: Dessert.
Origin: Island of Fyn, Denmark, 1910.
Parentage: ?Cox's Orange Pippin (female) x Unknown.
Flowering: Mid-season.

The flattened fruits, around 7cm (2¾in) across, are yellow with a pinkish red flush and some spotting. The creamy white flesh is firm, crisp, fine-textured and juicy with a fair flavour.

Note: Fruits can crack during some seasons. They are best eaten straight from the tree. Trees are moderately vigorous.

Jacquin

The conical fruits, around 7cm (2¾in) across, are bright yellowish green. The creamy white flesh is firm and crisp with a moderately sweet flavour.

Type: Dessert.
Origin: Boisbunel, France, recorded 1872.
Parentage: Unknown.
Flowering: Mid-season.

Note: Fruits are suitable for juicing and cider making. Trees are moderately vigorous.

Right: Jacquin is local to the Meurthe-et-Moselle department in the Lorraine region of France.

James Grieve

Left: James Grieve is a famous apple, formerly grown widely throughout Europe.

The rounded fruits, up to around 7cm (2¾in) across, are bright yellow-green with a bright orange-red flush and some streaking and spotting. The creamy white flesh is rather soft but very juicy with a good refreshing flavour.

Type: Culinary and dessert.
Origin: Edinburgh, Scotland, UK, first recorded 1893.
Parentage: Pott's Seedling or Cox's Orange Pippin (female) x Unknown.
Flowering: Mid-season.
Other names: Dzems Griw, Grieve and Jems Griv.

Note: Trees crop heavily, thriving in cool areas. Fruits do not store well. Bitter pit can be a problem. Commercially, this variety is often used for juicing and cider making. Red James Grieve is a sport.

Jansen von Welten

The very irregular fruits, around 8cm (3in) across, are yellowish green with a strong red flush and some spotting. The creamy white flesh is fairly firm with a sweet, aromatic, nutty flavour.

Note: Trees, of moderate vigour, crop freely. They are best trained as pyramids.

Type: Dessert.
Origin: Welten, nr Aachen, Germany, 1823.
Parentage: Unknown.
Flowering: Mid-season.
Other names: Couronne des Pommes, Fausen of Wellen, Jansen de Welten, Jansen van Welten, Jansen von Wetton, Reinette von Welten, Rosen Apfel von Welten and Rosenapfel von Welten.

Jean Tondeur

Type: Dessert.
Origin: France, 1947.
Parentage: Unknown.
Flowering: Late.

The flattened fruits, around 7cm (2¾in) across, are bright greenish yellow with a red flush. The white flesh is soft and fine with a subacid flavour.

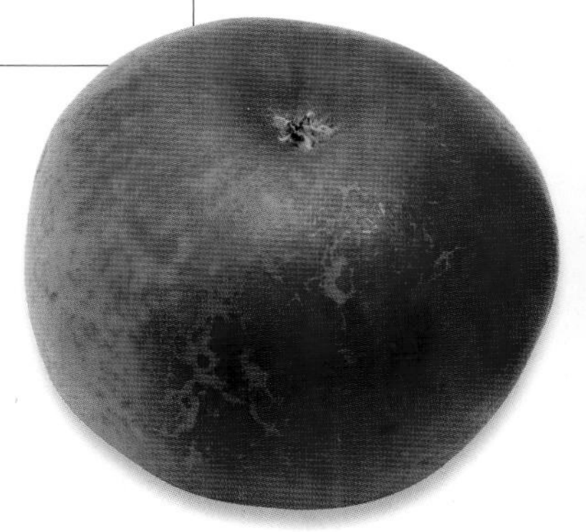

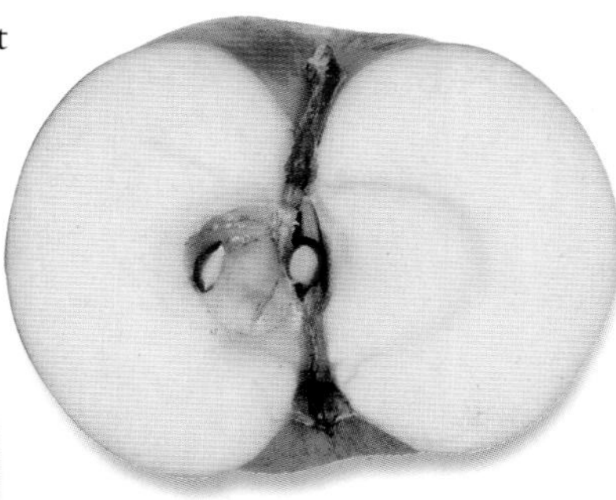

Note: This variety is local to the Marne department of north-eastern France. Trees are moderately vigorous.

Jersey Black

Type: Dessert.
Origin: New Jersey, USA, recorded 1817.
Parentage: Unknown.
Flowering: Mid-season.
Other names: Black American, Black Apple, Black Apple of America, Black Jersey, Dodge's Black and Small Black.

The uneven, very knobbly fruits, to 9cm (3¼in) across, are yellowish green with a dark red flush. The creamy-white flesh is crisp and rather coarse with a sweet, subacid, aromatic flavour.

Note: This variety is self-sterile, so a compatible pollination partner is required for successful fruiting.

Jonagored

Type: Dessert.
Origin: Halen, Belgium, introduced 1985.
Parentage: Golden Delicious (female) x Jonathan (male).
Flowering: Very early.

Note: Early flowering makes this triploid variety suitable only for areas where late frosts are rare.

The knobbly, slightly conical fruits, around 7cm (2¾in) across, are yellow-green with a strong dark red flush that can be patchy and some striping. The creamy white flesh is fine-textured and juicy with a sweet and rich flavour.

***Above**: This variety is a sport mutation of Jonagold.*

Jonathan

The irregular fruits, to 8cm (3in) across, are bright greenish-yellow with a strong bright red flush. The creamy-white flesh is soft, fine-textured, fairly juicy and sweet.

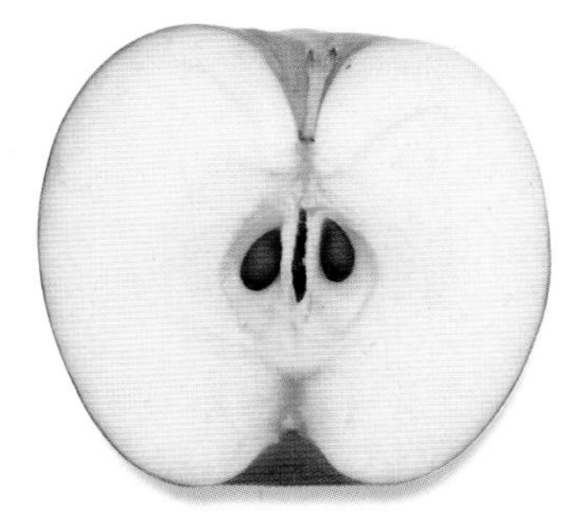

Note: Trees, which are naturally small, can be susceptible to fungal diseases. It is a popular variety in North America. It is partially self-fertile but best with a pollinator. The fruits store well.

Type: Dessert.
Origin: Woodstock, Ulster Co., NY, USA.
Parentage: ?Esopus Spitzenburg (female) x Unknown.
Flowering: Mid-season.
Other names: Djonathan, Dzhonatan, Dzonetn, Esopus Spitzenberg (New), King Philip, Philip Rick, Pomme Jonathan and Ulster Seedling.

Far left: This American variety is named after Jonathan Hasbrouck, who discovered the apple.

Josephine

The more or less rounded but uneven fruits, around 7cm (2¾in) across, are bright green with net-like russeting over the skin. The cream flesh is fine with a somewhat sweet flavour.

Note: This variety should not be confused with Belle Joséphine (correctly, Gloria Mundi). Trees are very vigorous.

Type: Dessert.
Origin: France, 1947.
Parentage: Unknown.
Flowering: Mid-season.

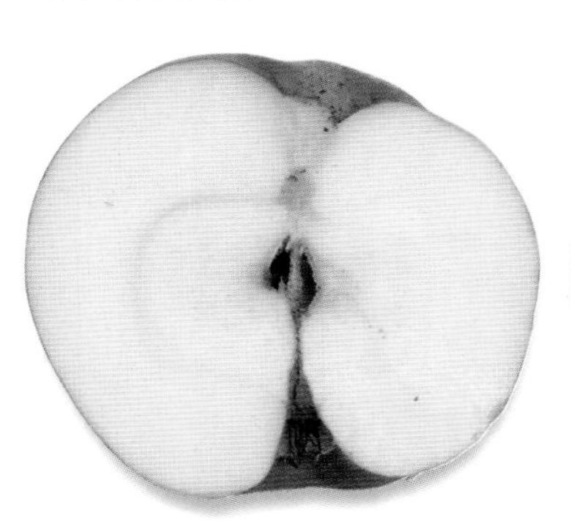

Joybells

The flattened, irregular fruits, around 7cm (2¾in) across, are pale greenish yellow with a strong pinkish red flush and some russeting. The creamy white flesh is crisp and juicy with a sweet and pleasant flavour.

Type: Dessert.
Origin: Godalming, Surrey, UK; records show that trees were grafted *c.*1914.
Parentage: Unknown.
Flowering: Mid-season.
Other name: Joy Bells.

Note: Trees, which crop heavily, are moderately vigorous.

Far left: Its pretty markings make Joybells a particularly attractive apple.

Jumbo Ohrin

Type: Dessert.
Origin: Nakajima Tenkohen Fruit Nurseries, Japan, 1985.
Parentage: ?Golden Delicious (female) x ?Indo (male).
Flowering: Early.

The rather uneven, somewhat square fruits, around 7cm (2¾in) across, are bright green with a red flush. The creamy white flesh is crisp and juicy.

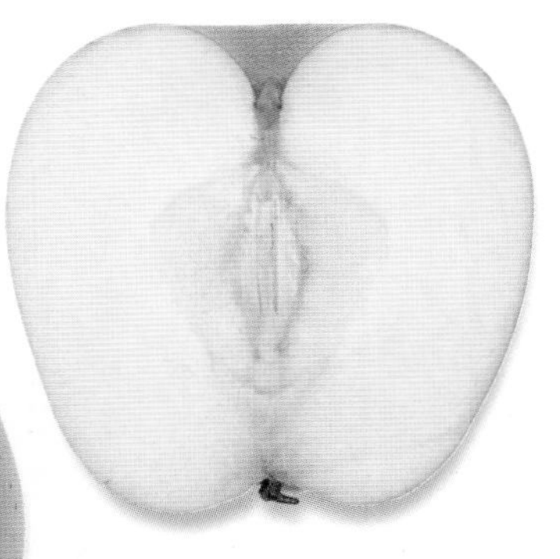

Note: Jumbo Ohrin is one of the most important commercial apple varieties in Japan.

Jupp's Russet A

Type: Dessert.
Origin: New Zealand, 1951.
Parentage: Unknown.
Flowering: Mid-season.

The rounded fruits, around 7cm (2¾in) across, are yellowish green with a light red flush and some russeting. The greenish white flesh is firm and fine with a sweet subacid flavour.

Note: Though it commemorates an English grower, this has been a valued variety in New Zealand.

Left: Russeting is a characteristic of this apple, as its name suggests.

Kandile

Type: Dessert.
Origin: Bulgaria, 1957.
Parentage: Unknown.
Flowering: Mid-season.

The slightly flattened fruits, around 7cm (2¾in) across, are bright greenish yellow with a red flush and striping. The creamy white flesh is firm and fine with a subacid flavour.

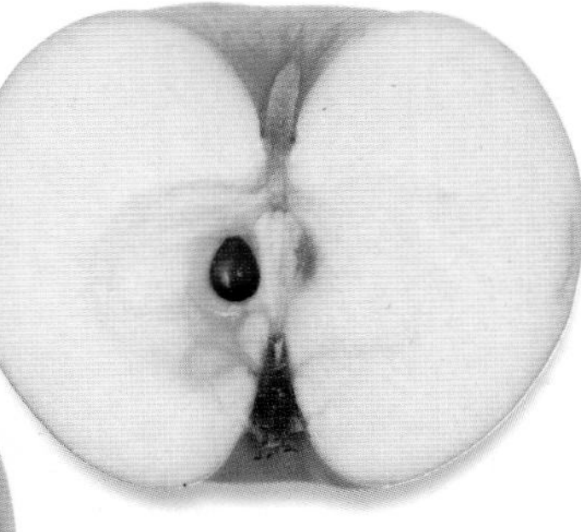

Note: This variety should not be confused with a much older variety known as Kandil Sinap, which is also sometimes grown as Kandile.

Kenneth

The rounded, occasionally irregular fruits, around 7cm (2¾in) across, are dull green with a dark red flush. The creamy white flesh is somewhat coarse and soft with a subacid and sweet flavour.

Type: Dessert.
Origin: Rhyl, Wales, UK, 1920.
Parentage: Unknown.
Flowering: Mid-season.

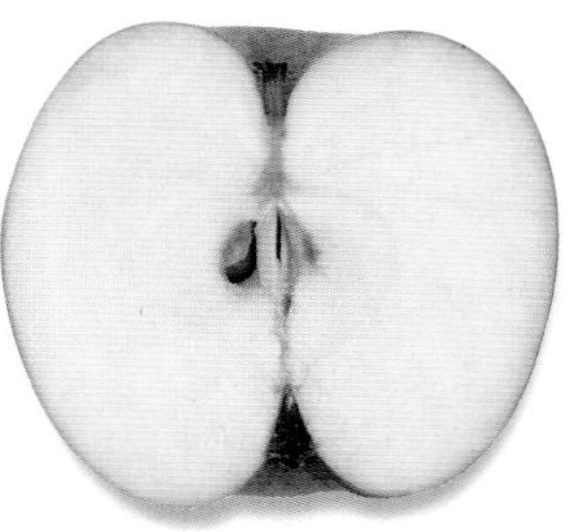

Note: This variety is named in honour of its breeder, Kenneth McCreadie. Trees are moderately vigorous.

Right: Fruits of this variety can be variously shaped and of different sizes.

Keswick Codlin

The conical fruits, around 7cm (2¾in) across or more, are clear bright green with an orange flush and some spotting. The yellowish white flesh is soft, rather coarse-textured and with an acid flavour.

Note: This is an excellent garden variety that crops prolifically. The flesh is very juicy when freshly picked but becomes dry and mealy after a few weeks' storage. Fruits cook to a sweet purée.

Left: Keswick Codlin is an easy apple variety to grow.

Type: Culinary.
Origin: Found on a rubbish heap at Gleaston Castle near Ulverston, Lancashire, UK, recorded 1793.
Parentage: Unknown.
Flowering: Early.
Other names: Codlin de Keswick, Everbearing, Keswick, Keswick Codling, Keswicker Kuchenapfel, Kodlin kesvikskii, Pinder's Apple and White Codlin.

Kidd's Orange Red

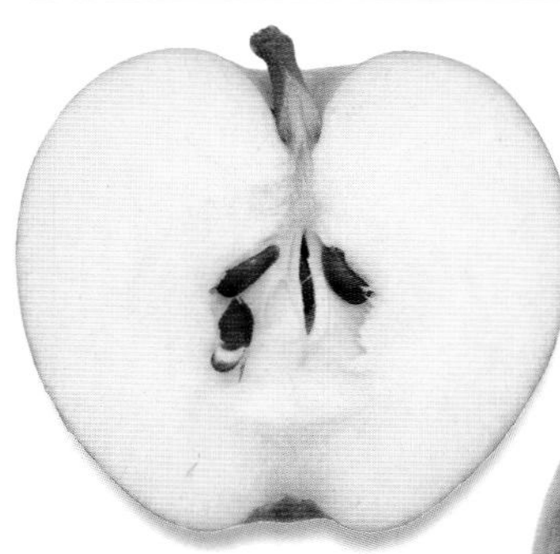

The rounded fruits (sometimes lumpy), around 7cm (2¾in) across, are yellow with a bright orange-red flush somewhat broken by russeting. The creamy white flesh is firm, crisp and juicy with a sweet, rich, aromatic flavour.

Type: Dessert.
Origin: Greytown, Wairarapa, New Zealand, 1924.
Parentage: Cox's Orange Pippin (female) x Delicious (male).
Flowering: Mid-season.
Other names: Delco, Kidd's Orange and Kidd's Oranje Roode.

Note: Trees crop reliably but seldom heavily – hence this variety has not become popular commercially. They benefit from thinning.

King David

Type: Dessert.
Origin: Washington County, AR, USA, found 1893 in a hedgerow.
Parentage: ?Jonathan (female) x Winesap or Arkansas Black (male).
Flowering: Mid-season.

The rounded but uneven fruits, around 7cm (2¾in) across, are yellow-green with a dark red flush and some striping. The creamy white flesh is rather coarse with a subacid, slightly sweet flavour.

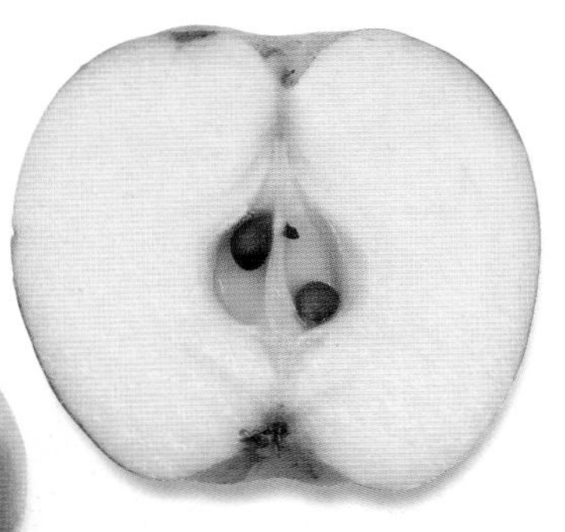

Note: Trees, ultimately large, bear very early and are resistant to fireblight. The fruits store well.

King of the Pippins

Type: Dessert.
Origin: UK or France, first recorded 1800.
Parentage: Unknown.
Flowering: Mid-season.
Other names: Aranyparmen, Pearmain, English Winter Golden Pear, George I, Gold Parmane, Goldreinette, Hampshire Yellow, Herzogs Reinette, Jones' Southampton Yellow, King of Pippins, Orange Pearmain, Pike's Pearmain, Reinette d'Oree, Seek no Farther, Shropshire Pippin and Winter Pearmain.

The rounded fruits, around 7cm (2¾in) across, are bright greenish yellow with a strong red flush. The white flesh is firm and juicy with a sharp and slightly aromatic subacid flavour.

Note: Formerly Golden Winter Pearmain. Identical to Reine des Reinettes grown in France. Trees do not do well in cold, heavy soils. It is partially self-fertile.

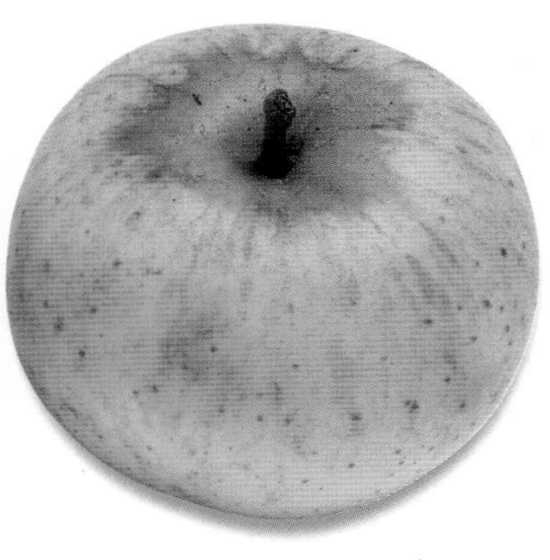

King's Acre Pippin

Type: Dessert.
Origin: King's Acre Nurseries, Hereford, recorded 1897, introduced 1899.
Parentage: ?Sturmer Pippin (female) x ?Ribston Pippin (male).
Flowering: Mid-season.
Other names: Cranston's, Cranston's Pippin and Ribston Pearmain.

The rounded fruits, up to around 6cm (2¼in) across or more, are yellow-green with a red flush and some russeting. The creamy white flesh is firm, coarse-textured and juicy with a rich, strong, aromatic flavour.

Note: The fruits store well for one to two months. This variety is a triploid. Bitter pit can be a problem.

Kis Erno Tabornok

Type: Dessert.
Origin: Hungary, 1948.
Parentage: Unknown.
Flowering: Mid-season.

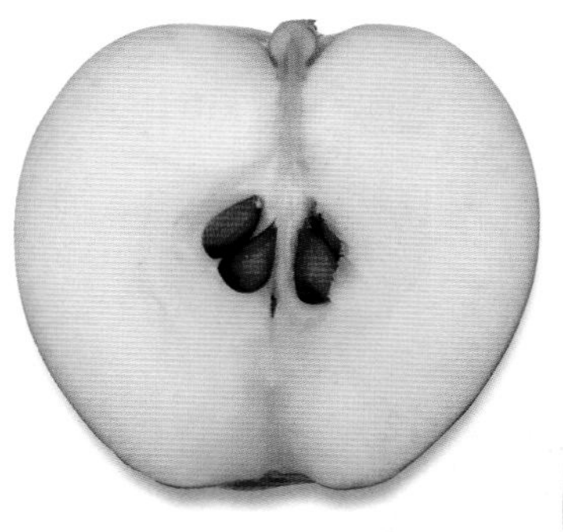

Note: This unusual apple is not widely grown outside its native Hungary.

Left: This variety honours a Hungarian member of the military.

The somewhat flattened, uneven fruits, around 7cm (2¾in) across, are bright green with a red flush that can show as patches. The creamy-white flesh is tough with a slightly sweet flavour.

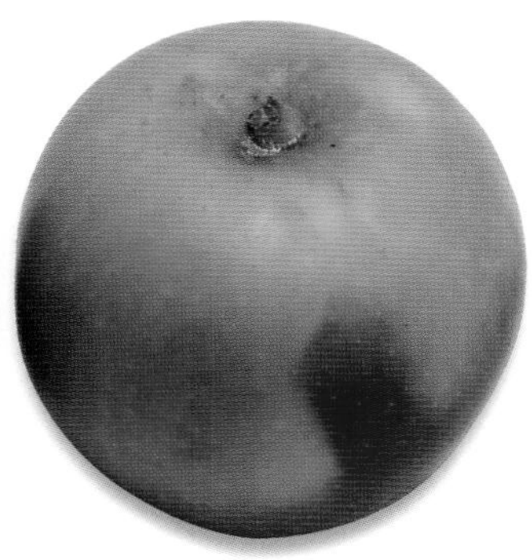

Kitchovka

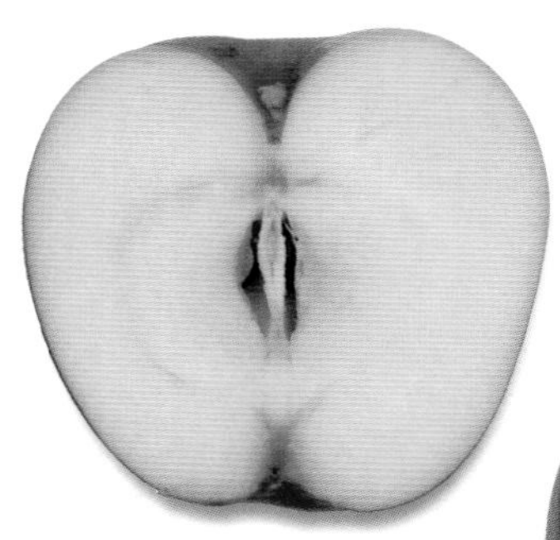

Note: Trees are very vigorous.

The slightly irregular fruits, around 6cm (2¼in) across, are yellow with a pinkish red flush. The creamy white flesh is firm and rather coarse with a subacid and slightly sweet flavour.

Type: Dessert.
Origin: Bulgaria, 1957.
Parentage: Unknown.
Flowering: Mid-season.

Above: Fruits within a cluster are of uneven shape.

Klunster

The flattened, slightly irregular fruits, around 7cm (2¾in) across, are bright green, sometimes with a reddish flush and some russeting. The white flesh is firm and fine with a subacid flavour.

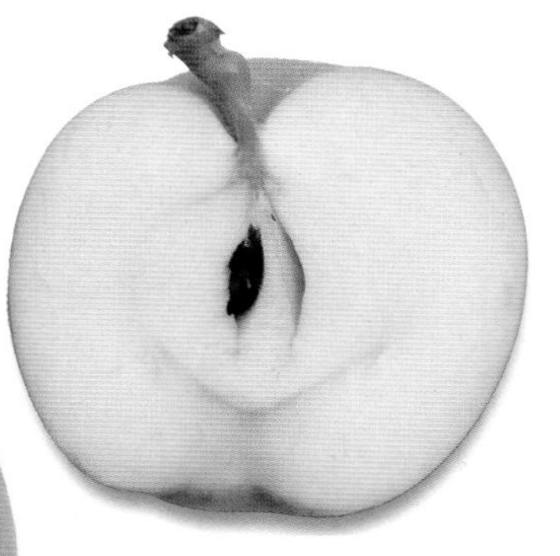

Note: Trees are very vigorous.

Right: This variety generally crops well.

Type: Dessert.
Origin: Jork Fruit Research Station, nr Hamburg, Germany, 1951.
Parentage: Unknown.
Flowering: Mid-season.

Knobby Russet

The irregular fruits, around 7cm (2¾in) across, have a distinctively warty appearance and are dull yellow-green with some scaly russeting. The white flesh is firm and rather dry with a fairly strong flavour.

Note: This variety is self-sterile so a suitable pollination partner is needed for good fruiting. Trees are moderately vigorous.

Type: Dessert.
Origin: Midhurst, Sussex, UK, 1820.
Parentage: Unknown.
Flowering: Early.
Other names: Knobbed Russet, Old Maid's, Old Maids, Winter, Winter Apple and Winter Russet.

Kolacara

Type: Dessert.
Origin: Former Yugoslavia, 1936.
Parentage: Unknown.
Flowering: Early.
Other names: Kolatchara and Koltchara.

The flattened, somewhat uneven fruits, around 7cm (2¾in) across or more, are greenish yellow with a bright red flush and striping. The creamy white flesh is firm and coarse with a subacid flavour.

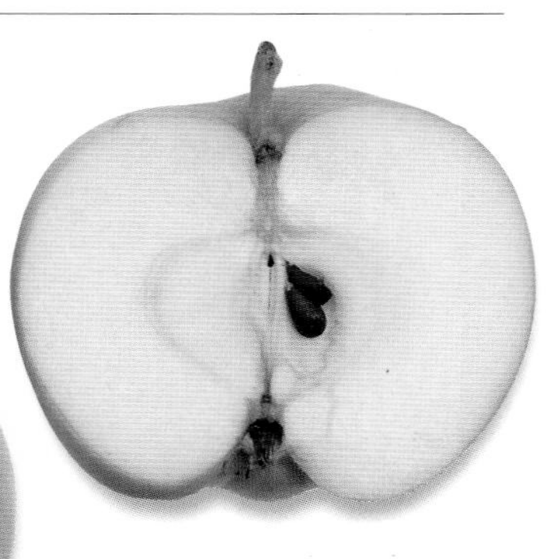

Note: This variety is a triploid so needs two pollination partners. Trees are very vigorous.

Kougetsu

Type: Dessert.
Origin: Aomori Apple Experimental Station, Japan, date unknown.
Parentage: Golden Delicious (female) x Jonathan (male).
Flowering: Very early.

The rounded, slightly uneven fruits, around 7cm (2¾in) across, are bright yellowish green with a bright red flush and some spotting. The cream flesh is coarse in texture.

Note: Early flowering makes this variety suitable only for areas where late frosts are rare.

La Gaillarde

The rounded fruits, around 6cm (2¼in) across, are green with a red flush. The creamy white flesh is coarse and crisp with a subacid sweet flavour.

Note: This variety is self-sterile so a suitable pollination partner is needed for good fruiting. Trees are very vigorous.

Type: Culinary and dessert.
Origin: Angers, France, 1930.
Parentage: Unknown.
Flowering: Mid-season.

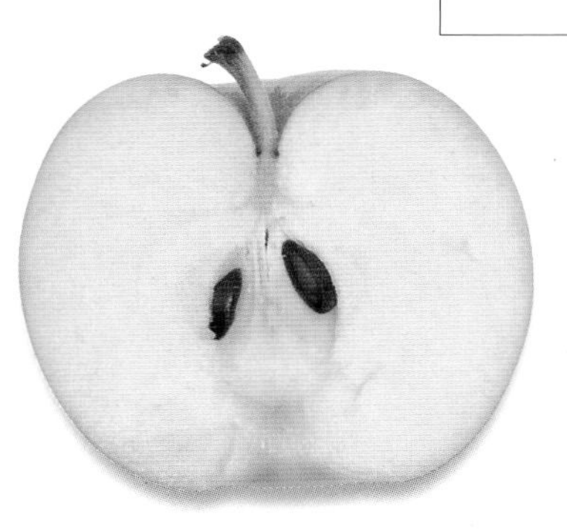

La Nationale

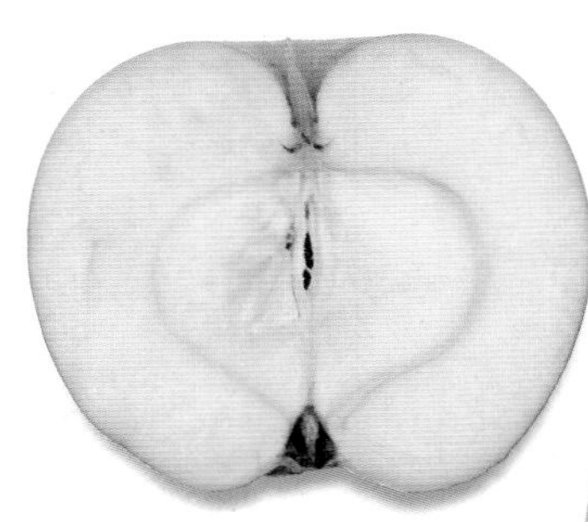

The rounded fruits, around 7cm (2¾in) across, are green with a red flush. The whitish flesh is firm with a sweet flavour.

Type: Dessert.
Origin: Sainte-Romain-au-Mont d'Or, Rhône, France, 1871.
Parentage: Unknown.
Flowering: Late.
Other names: Bernardin, Bourget, Cusset à Fruits Rouges, Cusset Rouge, Déesse nationale and Natsionalnoe.

Note: Late flowering makes this a suitable variety for cold areas. Fruits can also be used for cooking.

Lady Henniker

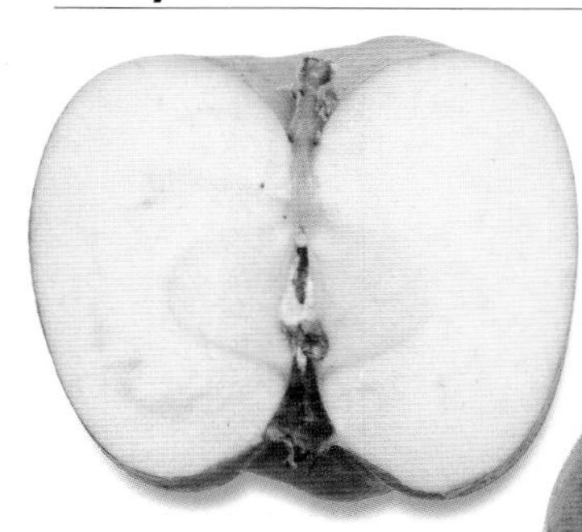

The rather knobbly, somewhat flattened and irregular fruits, around 7cm (2¾in) across, are light green with a red flush that can show as flecking. The whitish flesh is firm, coarse-textured and rather dry with a fairly acid but fair flavour.

Type: Culinary and dessert.
Origin: Thornham Hall, Eye, Suffolk, UK, between 1840 and 1850, introduced 1873.
Parentage: Unknown.
Flowering: Mid-season.
Other names: Henniker, Lady Hennicker and Ledi Genniker.

Note: This popular garden variety is self-sterile so a suitable pollination partner is needed for good fruiting. Trees are very vigorous.

Lady Sudeley

Type: Dessert.
Origin: Petworth, Sussex, UK, *c.*1849, introduced 1885.
Parentage: Unknown.
Flowering: Mid-season.
Other names: Jacob's Strawberry and Lady Sudely.

The rounded, slightly irregular, ribbed fruits, around 7cm (2¾in) across, are bright yellow-green with a pinkish red flush and some streaking and russeting. The creamy yellow flesh is firm and juicy with a somewhat acid but good flavour.

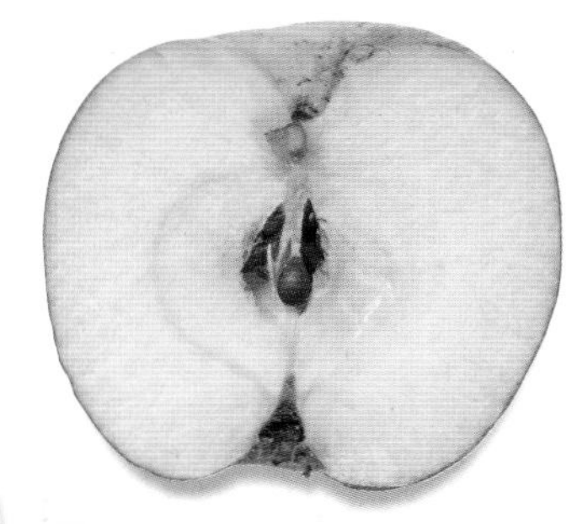

Note: Trees are small, compact and moderately vigorous. A tip-bearer.

Lady Williams

Type: Dessert.
Origin: Donnybrook, Western Australia, *c.*1935.
Parentage: ?Granny Smith (female) x Jonathan or ?Rokewood (male).
Flowering: Early.

Note: Fruits need a long summer to ripen fully. Trees are moderately vigorous.

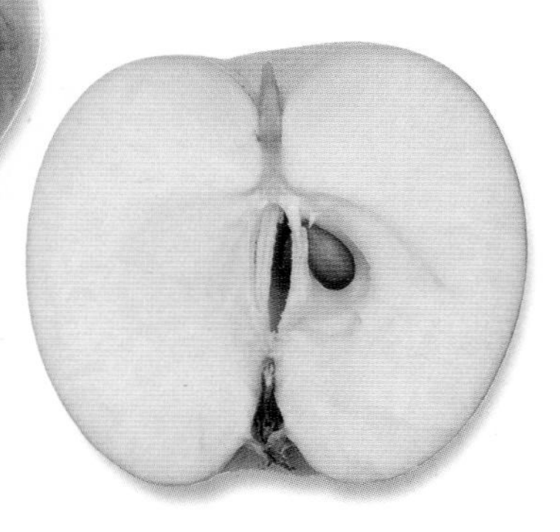

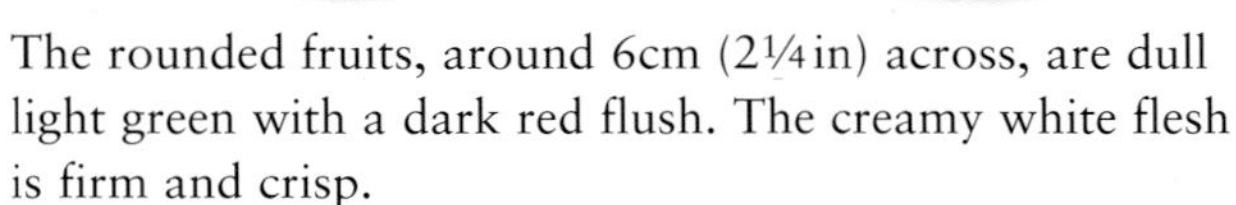

The rounded fruits, around 6cm (2¼in) across, are dull light green with a dark red flush. The creamy white flesh is firm and crisp.

Landsberger Reinette

Type: Dessert.
Origin: Landsberg/Warthe, Brandenburg, Germany, *c.*1840.
Parentage: Unknown.
Flowering: Early.
Other names: Buchardt Renette, Landsberg, Landsberger, Landsberger Renette, Landsberska, Landsberska Reneta, Reinette de Landsberg, Renet landsbergskii, Reneta Gorzawska, Reneta Landsberska and Surprise.

The rounded fruits, around 7cm (2¾in) across, are bright green. The whitish flesh is soft, fine-textured and very juicy with a sweet and refreshing flavour.

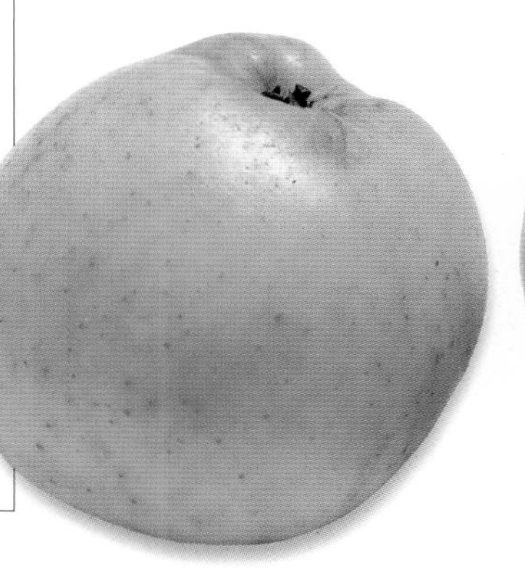

Note: Fruits store well. They are also suitable for cooking.

Lane's Prince Albert

Type: Culinary.
Origin: Berkhamstead, Hertfordshire, UK, *c.*1840, introduced 1850.
Parentage: Russet Nonpareil (female) x Dumelow's Seedling (male).
Flowering: Mid-season.
Other names: Albert Lanskii, Lane, Lane's, Lane's Albert, Perkins' A. 1, Prince Albert, Prince Albert de Lane, Prinz Albert, Profit, Victoria and Albert.

The slightly uneven fruits, more than 7cm (2¾in) across, are bright green with a red flush and some flecking. The creamy white flesh is very juicy with an acid flavour.

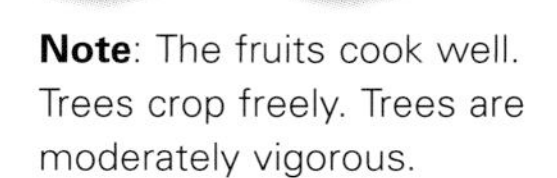

Note: The fruits cook well. Trees crop freely. Trees are moderately vigorous.

Lawyer Nutmeg

Note: This variety originated with David Lawyer. Trees are moderately vigorous.

The slightly flattened, somewhat uneven fruits, around 7cm (2¾in) across, are bright yellow green with a slight red flush. The creamy white flesh has a rich and spicy flavour reminiscent of nutmeg.

Type: Dessert.
Origin: Plains, MT, USA, date unknown.
Parentage: ?Wismer Dessert (female) x ?Apple Crab (male).
Flowering: Mid-season.

Laxton's Royalty

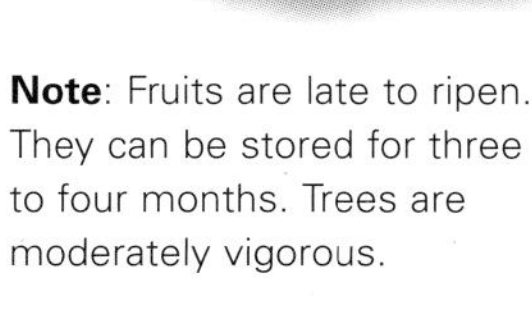

Note: Fruits are late to ripen. They can be stored for three to four months. Trees are moderately vigorous.

The somewhat flattened fruits, to 8cm (3in) across, are mid-green with a dark red flush. The yellowish white flesh is hard and crisp with a slightly sweet to subacid taste.

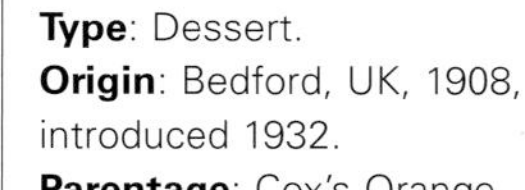

Type: Dessert.
Origin: Bedford, UK, 1908, introduced 1932.
Parentage: Cox's Orange Pippin (female) x Court Pendu Plat (male).
Flowering: Late.

Laxton's Superb

The rounded, slightly uneven fruits, around 7cm (2¾in) across, are dull green with a dull red flush. The creamy white flesh is firm and very juicy with a sweet, pleasant and refreshing flavour.

Note: Trees, which are vigorous and spreading, are prone to biennial bearing.

Type: Dessert.
Origin: Bedford, UK, 1897, introduced 1922.
Parentage: Wyken Pippin (female) x Cox's Orange Pippin (male).
Flowering: Mid-season.
Other names: Laxton Superb, Laxtons Superb and Superb.

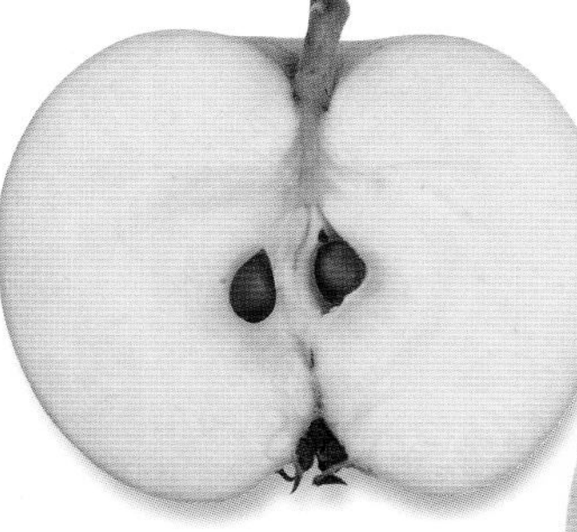

Leather Jacket

Type: Culinary.
Origin: Harwich, Essex, UK, 1883.
Parentage: Unknown.
Flowering: Early.

The somewhat flattened fruits, around 7cm (2¾in) across, are bright yellow-green with some spotting. The creamy white flesh is somewhat acid in flavour.

Note: The fruits cook well.

Left: The skins of this variety have a rather greasy appearance and feel.

Leathercoat Russet

Type: Dessert.
Origin: England, UK, first recorded 1597.
Parentage: Unknown.
Flowering: Early.
Other name: Royal Russet.

The irregular, somewhat flattened fruits, around 7cm (2¾in) across, are dull green with some russeting. The greenish yellow flesh is tender with a sweet, subacid flavour.

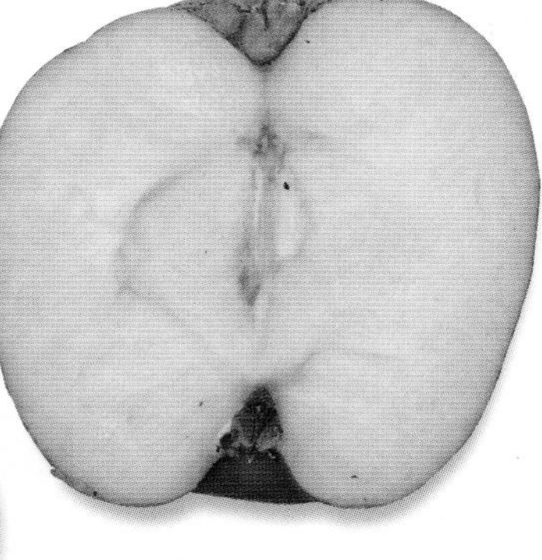

Note: This is a very vigorous variety that is best grown on a dwarfing rootstock.

Legana

The rounded, slightly conical fruits, around 6cm (2¼in) across, are yellowish green with a red flush. The creamy white flesh is firm and fine with a sweet, subacid flavour.

Type: Dessert.
Origin: Legana, Tasmania, Australia, *c.*1940.
Parentage: Democrat (female) x Delicious (male).
Flowering: Mid-season.

Note: This variety is weak-growing so may not do well on a dwarfing rootstock.

Right: Legana is named after a rural town that contained several apple orchards.

Leonie de Sonnaville

The flattened fruits, around 6cm (2¼in) across, are dull greenish yellow with a strong red flush. The creamy white flesh is sweet and juicy with an aromatic flavour.

Type: Dessert.
Origin: Institute for Horticultural Plant Breeding (IVT), Wageningen, Netherlands, 1974.
Parentage: Cox's Orange Pippin (female) x Jonathan (male).
Flowering: Late.

Note: This variety is a partial tip-bearer. Trees are moderately vigorous.

Levering Limbertwig

The slightly flattened fruits, around 6cm (2¼in) across, are bright green with a strong red flush. The yellowish flesh is fairly firm with a subacid flavour.

Type: Dessert.
Origin: Ararat, VA, USA, date unknown.
Parentage: Unknown.
Flowering: Mid-season.
Other name: Limbertwig.

Note: A larger-fruited mutation of Limbertwig. It is very vigorous, so is suitable for growing on a dwarfing rootstock. Fruits ripen late.

Lille

Type: Dessert.
Origin: France, 1948.
Parentage: Unknown.
Flowering: Late.

The rounded fruits, around 5cm (2in) across, are bright green with a deep pinkish red flush. The greenish white flesh is fine with a subacid flavour.

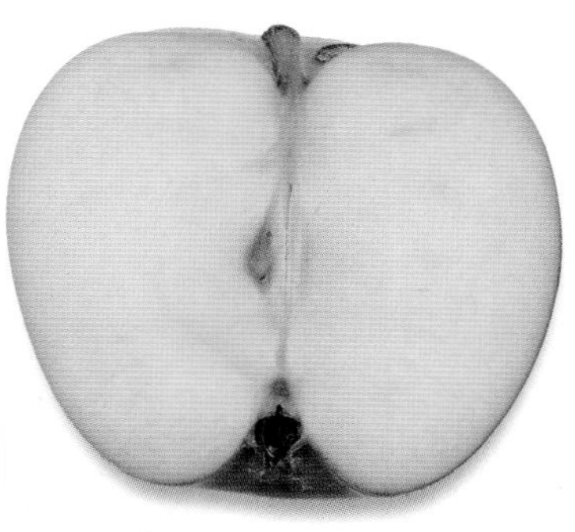

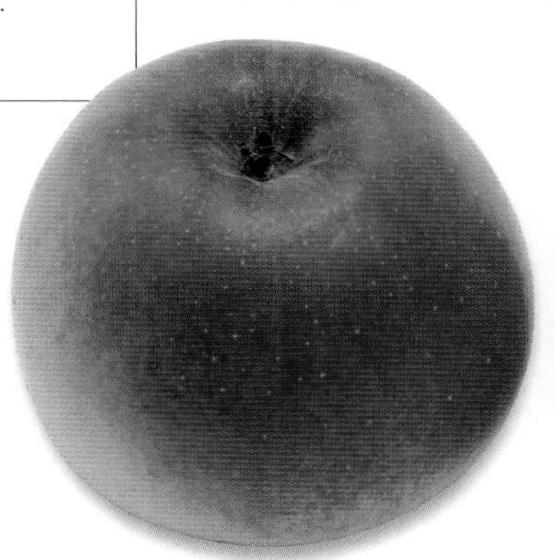

Note: Late flowering makes this a suitable variety for growing in areas where late spring frosts are likely.

Lodgemore Nonpareil

Type: Dessert.
Origin: Lodgemore, Stroud, Gloucestershire, UK, *c.*1808.
Parentage: Unknown.
Flowering: Late.
Other names: Clissold's Seedling, Lodgemore Seedling, Non Pareille de Lodgemore and Nonpareille de Lodgemore.

The somewhat irregular fruits, around 6cm (2¼in) across, are bright yellow-green, dotted with grey and with some russeting. The creamy white flesh is firm, crisp and juicy with a sweet and perfumed flavour.

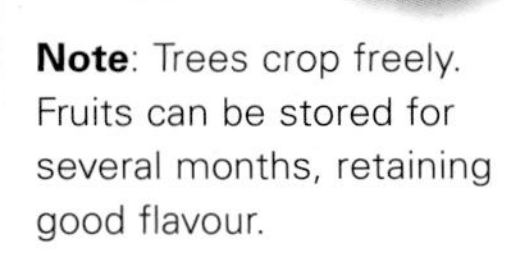

Note: Trees crop freely. Fruits can be stored for several months, retaining good flavour.

Lombarts Calville

Type: Dessert.
Origin: Zundert, Netherlands, 1906.
Parentage: ?Calville Blanc D'Hiver (female) x Unknown.
Flowering: Mid-season.
Other names: Lombarts Kalvill, Lombartscalville and Witte Winter Lombartscalville.

The slightly conical fruits, around 6cm (2¼in) across, are a shiny yellowish green. The creamy white flesh is firm, fairly coarse and soft with a sweet subacid flavour.

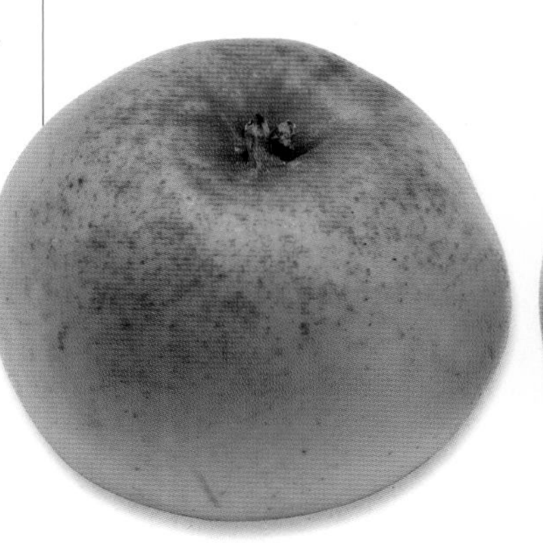

Note: Trees are moderately vigorous. Fruits can be stored for around three months.

Long Bider

Type: Culinary.
Origin: East Malling Research Station, Maidstone, Kent, UK, 1948.
Parentage: Unknown.
Flowering: Unknown.
Other name: Longbider.

The flattened fruits, around 6cm (2¼in) across, are yellowish green with some spotting. The flesh is creamy white.

Note: The fruits cook well.

Longney Russet

The rounded but slightly irregular fruits, to 7cm (2¾in) across, are green with heavy russeting and a red flush that can appear as spotting. The creamy white flesh is fine, hard and dry with a subacid flavour.

Note: The fruits are suitable for cider making. The fruits store well, retaining good flavour.

Type: Dessert.
Origin: Gloucestershire, UK, 1949 (probably older).
Parentage: Unknown.
Flowering: Mid-season.

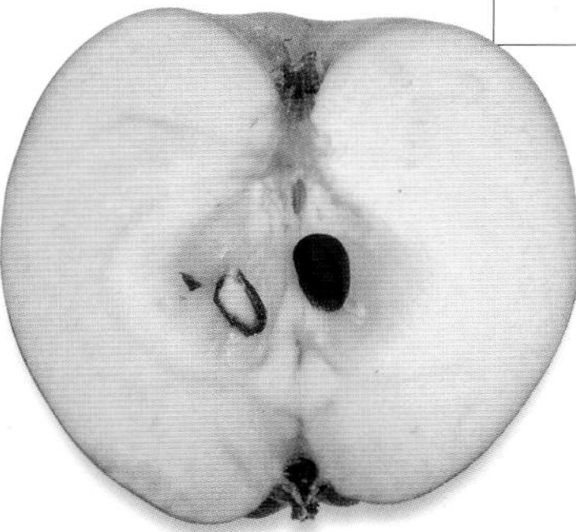

Lord Burghley

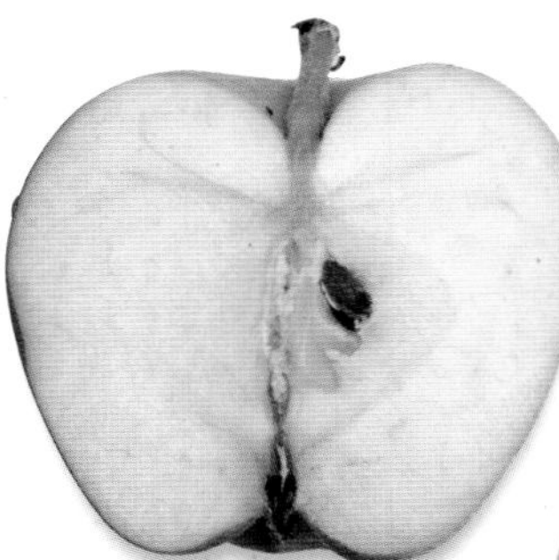

Note: The fruits keep well in storage, retaining good flavour for up to five months.

The slightly conical fruits, around 7cm (2¾in) across, are bright golden yellow-green with a strong red flush. The yellowish white flesh is very firm, fine-textured and rather dry with a sweet and rich aromatic flavour.

Type: Dessert.
Origin: Burghley, Stamford, Lincolnshire, UK, 1843, introduced 1865.
Parentage: Unknown.
Flowering: Late.
Other names: Bergli, Lord Burghleigh and Lord Burleigh.

Lord Grosvenor

Type: Culinary.
Origin: Unknown but believed to be new in 1872.
Parentage: Unknown.
Flowering: Mid-season.

The very irregular, sometimes quince- or pear-like fruits, around 8cm (3in) across or more, are bright yellow-green to straw yellow with a few spots and some traces of thin russeting. The white flesh is soft, tender and juicy with an acid to subacid flavour.

Note: The fruits cook well. They can be stored for two to three months.

Lord Hindlip

Type: Dessert.
Origin: Worcestershire, UK, 1896.
Parentage: Unknown.
Flowering: Mid-season.

The roughly conical, somewhat irregular fruits, around 7cm (2¾in) across, are yellow-green with some russeting and a strong red flush. The creamy white flesh is fairly firm, very fine-textured and juicy with a good aromatic flavour.

Note: Trees are moderately vigorous. This variety is self-sterile, so a suitable pollination partner is needed for good fruiting.

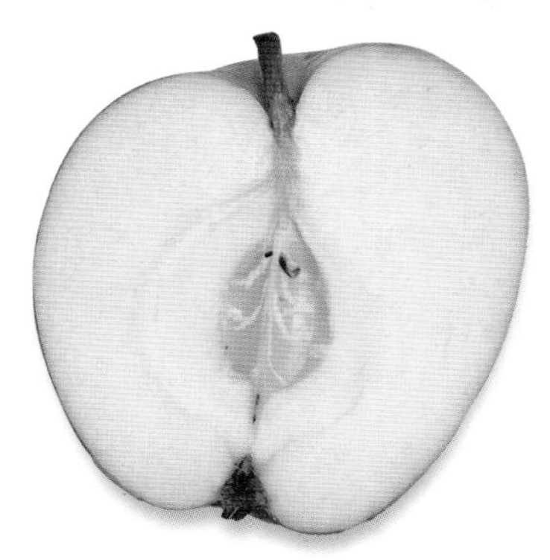

Lord Lambourne

Type: Dessert.
Origin: Bedford, UK, 1907, introduced 1923.
Parentage: James Grieve (female) x Worcester Pearmain (male).
Flowering: Early.

The slightly flattened but uniform fruits, around 6cm (2¼in) across, are bright yellow-green with a strong orange-red flush and some streaking. The creamy white flesh is slightly coarse-textured and juicy with a strong, sweet and somewhat aromatic flavour.

Above: Lord Lambourne is a classic dessert apple, with a uniform shape and a pretty flush.

Note: The same parents produced Katy and Elton Beauty. Trees, of moderate vigour, are easy to grow and crop well. They are partially self-fertile but fruit best in the presence of a pollinator.

Mabbot's Pearmain

Type: Dessert.
Origin: Kent, UK, first described 1883.
Parentage: Unknown.
Flowering: Mid-season.
Other names: Canterbury, Mabbut's Pearmain, Parmaene von Mabbott, Parmane von Mabot, Parmane von Mabott, Pearmain de Mabbot and Pearmain de Mabbott.

The rounded fruits, up to around 7cm (2¾in) across, are bright yellow-green with a pinkish red flush and some spotting and grey russeting. The yellowish white flesh is fairly crisp, somewhat coarse-textured and juicy with a slightly acid, pleasant aromatic flavour.

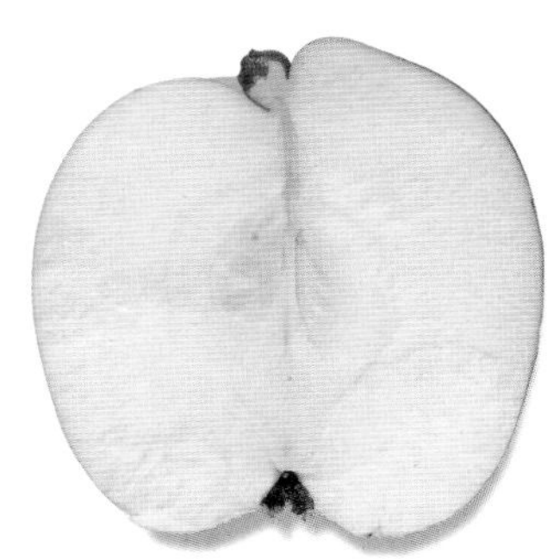

Above: Fruits can turn bright red on the side that is exposed to the sun.

Note: Fruits can be stored for up to around three months. Trees are moderately vigorous.

Macwood

The somewhat irregular, rounded to conical fruits, around 6cm (2¼in) across, are bright green with a strong dark red flush. The white flesh is firm and somewhat coarse with a sweet, perfumed flavour.

Type: Dessert.
Origin: Central Experimental Farm, Ottawa, Canada, 1936.
Parentage: McIntosh (female) x Forest (male).
Flowering: Early.

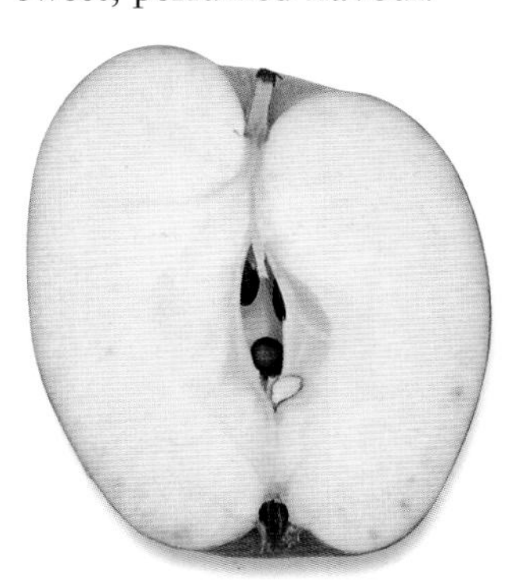

Note: Needs a sheltered position to protect the flowers in frost-prone areas.

Right: Dark red apples such as this one are always popular.

Madoue Rouge

The irregular fruits, around 6cm (2¼in) across, are yellow-green with a red flush. The yellowish flesh is tough with a sweet subacid flavour.

Type: Dessert.
Origin: Nieul (Haute-Vienne), France, described 1947.
Parentage: Unknown.
Flowering: Very late.
Other name: Madou.

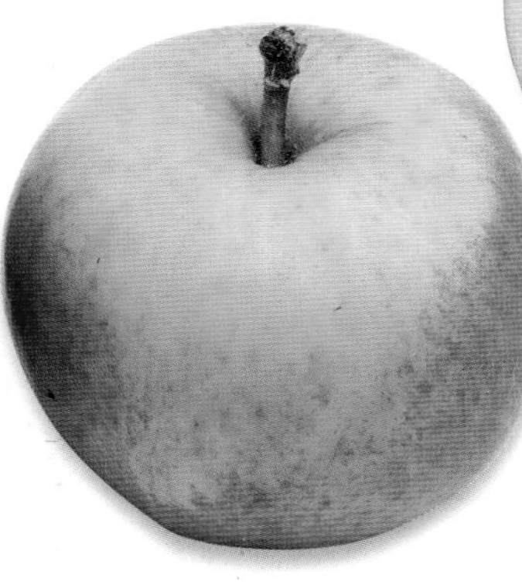

Note: Late flowering makes this a suitable variety for cold districts. It is local to the Limousin province of France.

Malling Kent

Type: Dessert.
Origin: East Malling Research Station, Maidstone, Kent, UK, 1949.
Parentage: Cox's Orange Pippin (female) x Jonathan (male).
Flowering: Mid-season.

The conical fruits, around 6cm (2¼in) across, are bright greenish yellow with a pinkish red flush and some streaking. The creamy white flesh is slightly coarse-textured and fairly juicy with a pleasant aromatic flavour.

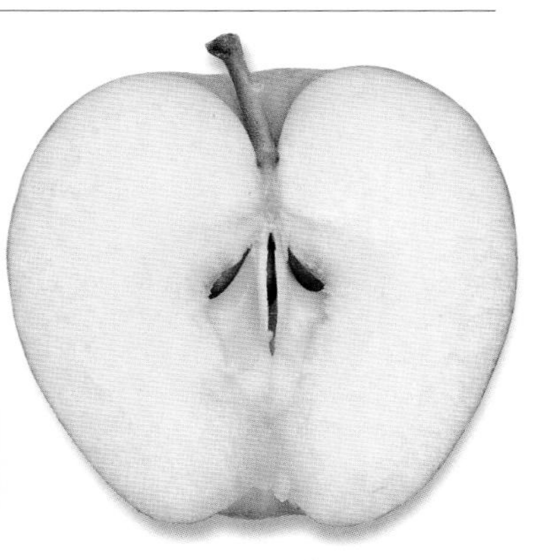

Note: Trees are of moderate vigour. The flavour is similar to that of Cox's Orange Pippin.

Mannington's Pearmain

Type: Dessert.
Origin: Uckfield, Sussex, UK, *c.*1770, introduced 1847.
Parentage: Unknown.
Flowering: Mid-season.
Other names: Mannington Pearmain, Mannington's Parmaene, Mannington's Parmane, Pearmain de Mannington and Pomme de Mannington.

The rounded to flattened fruits, around 7cm (2¾in) across, are light greenish yellow with thin russeting and a pinkish red flush. The flesh is greenish white to yellow, firm, fine-textured and moderately juicy with a slightly aromatic flavour.

Note: This variety was produced from seed found in cider pomace. To develop their fullest flavour, fruits are best allowed to hang on the tree for as long as possible before picking. Young trees bear well.

Margil

Type: Dessert.
Origin: Europe, known in England before 1750 (possibly of French origin).
Parentage: Unknown.
Flowering: Early.
Other names: Fail-me-Never, Gewurz Reinette, Herefordshire Margil, Kleine Granat-Reinette, Margil Hook, Munches Pippin, Muscadet, Never Fail, Reinette Muscat, Reinette Sucrée, Renet Muscat, Renetta Moscata, Small Ribston, Sucrée d'Hiver and White Margil.

The rounded to conical fruits, around 6cm (2¼in) across, are bright yellow-green with a strong orange-red flush and some streaking, with a tendency to russeting. The creamy white to yellow flesh is firm and rather dry with a sweet aromatic taste.

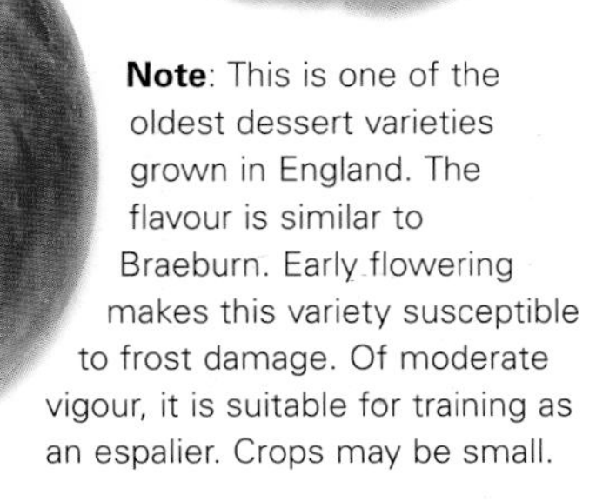

Note: This is one of the oldest dessert varieties grown in England. The flavour is similar to Braeburn. Early flowering makes this variety susceptible to frost damage. Of moderate vigour, it is suitable for training as an espalier. Crops may be small.

Mariborka

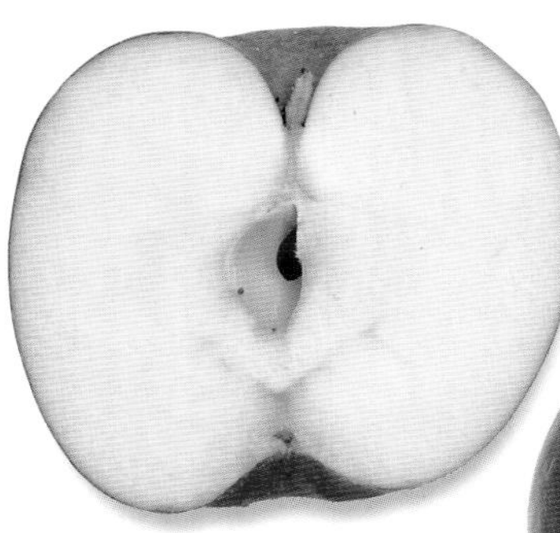

The somewhat flattened, slightly irregular fruits, around 6cm (2¼in) across, are bright green with a strong red flush and some light streaking. The whitish flesh is crisp and juicy with a rich, aromatic flavour.

Type: Dessert.
Origin: Institut Za Vocarstvo, Cacak, former Yugoslavia, date unknown.
Parentage: Golden Pearmain (female) x Jonathan (male).
Flowering: Late.

Note: Late flowering makes this a suitable variety for areas where late frosts are common.

Marie Doudou

The rather flattened, irregular fruits, around 6cm (2¼in) across, are dull pale green with a strong dull red flush and some russeting. The white flesh is tinged green and is fairly tough with a subacid flavour.

Type: Dessert.
Origin: France, described 1948.
Parentage: Unknown.
Flowering: Very late.

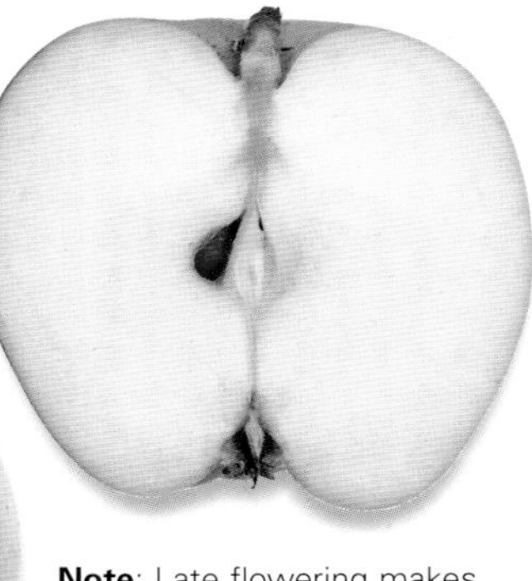

Note: Late flowering makes this a suitable variety for cold districts.

Right: Marie Doudou is local to north-western France.

Marie-Joseph d'Othée

The flattened fruits, to 5cm (2in) across, are bright greenish yellow with a pinkish red flush with some streaking and russeting. The greenish white flesh is firm with a sweet subacid flavour.

Type: Dessert.
Origin: Liège, Belgium, 1947.
Parentage: Unknown.
Flowering: Late.
Other names: de Fer, Ijzerappel, Marie-José, Pomme de Deux Ans, Pomme de Fer, Reine Marie d'Otlée and Reine Marie Joseph d'Othée.

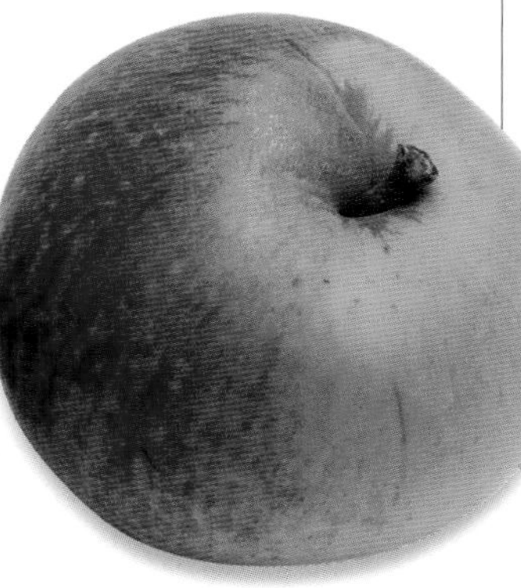

Note: This is a large tree, suitable for orchards, and generally healthy and productive. The fruits store well.

Marie-Louise Ducote

The rounded fruits, around 6cm (2¼in) across, are light yellow-green with some spotting and russeting. The greenish white flesh is tender with a sweet flavour.

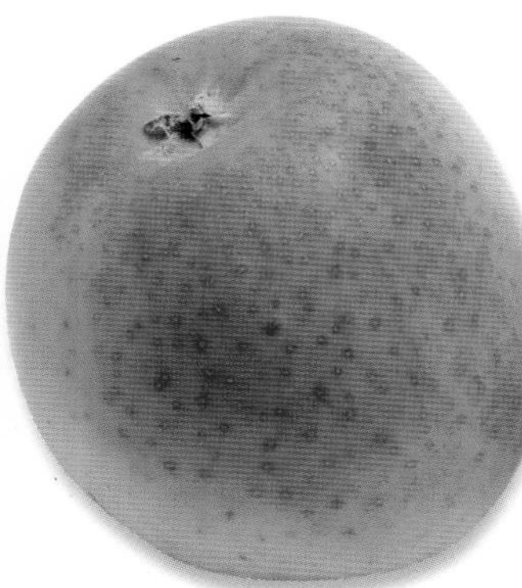

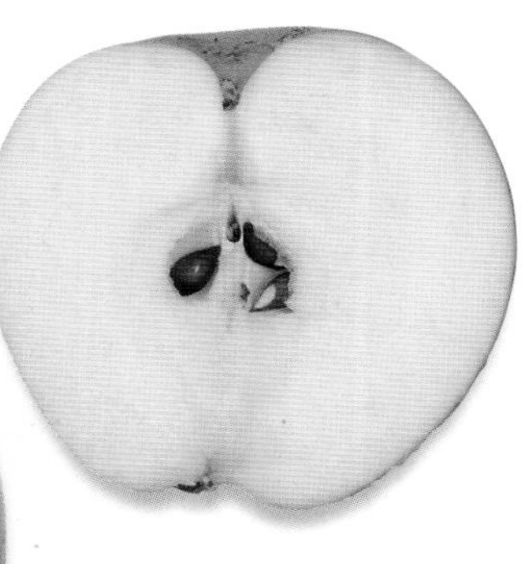

Type: Dessert.
Origin: France, 1947.
Parentage: Unknown.
Flowering: Late.

Note: This variety is spur-bearing.

Right: This dainty-looking apple has a very pleasant taste.

Marie-Madeleine

Type: Dessert.
Origin: France, 1947.
Parentage: Unknown.
Flowering: Late.
Other name: Marie Madeleine.

The rounded to conical fruits, around 6cm (2¼in) across, are pale green with russeting. The greenish white flesh is coarse, crisp and juicy with a subacid flavour.

Note: Besides being excellent as an eating apple, this variety is also good for cider making.

Marosszeki Piros Paris

Type: Dessert.
Origin: Maros-Torda, Hungary, first recorded in 1598.
Parentage: Unknown.
Flowering: Early.
Other names: Grosse-Pomme-Paris, Maros Szeki, Paris, Paris Alma, Paris Apfel, Paris de Moros, Paris jaune, Paris vert, Pomme Paris, Pomme Paris, Rouge de Marosszek and Rother Parisapfel.

The rounded fruits, around 6cm (2¼in) across, are bright yellow-green with a red flush. The white flesh has a subacid, aromatic flavour.

Note: This is an apple of great antiquity. Its place of origin is presently in central Romania (eastern Transylvania).

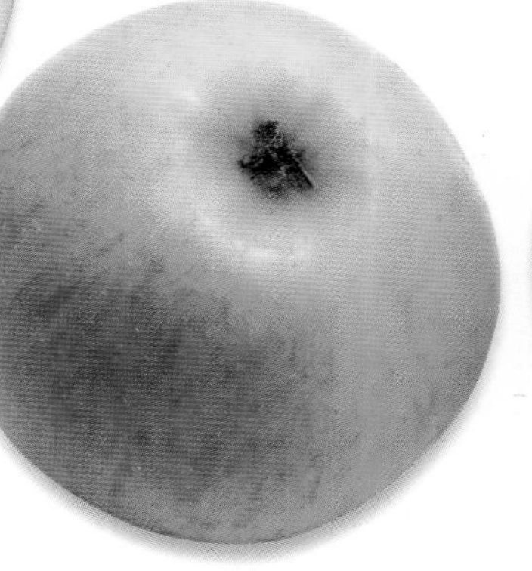

May Queen

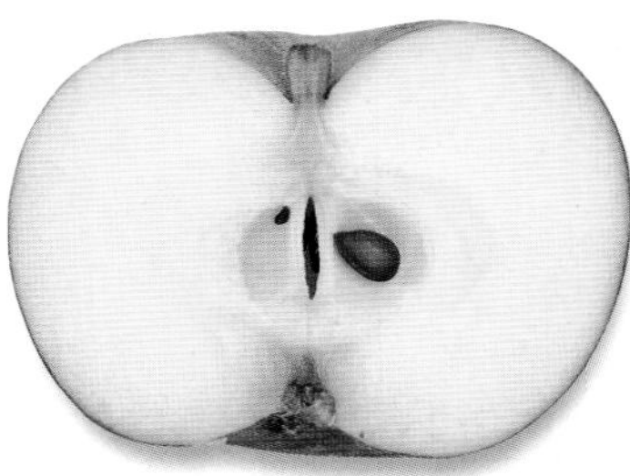

The flattened, irregular fruits, around 7cm (2¾in) across, are bright green with a deep red flush and some striping. The yellow flesh is firm and compact with an unusual, sometimes astringent flavour.

Type: Dessert.
Origin: Worcester, UK, 1888.
Parentage: Unknown.
Flowering: Mid-season.

Note: This variety is self-sterile, so a suitable pollination partner is required for successful fruiting. The fruits store well.

Right: *May Queen has a notable crunchy texture.*

McIntosh

Note: Trees, of moderate vigour, can be vulnerable to canker in humid areas. They are resistant to powdery mildew and cedar apple rust. Fruits store well but can lose flavour. This variety has given rise to several sports.

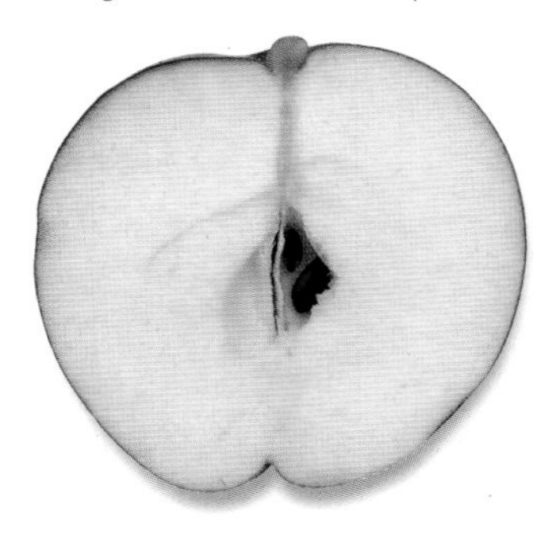

The rounded, often irregular fruits, around 7cm (2¾in) across, are bright green with a strong dark bluish red flush and some striping and flecking. The white flesh is rather soft, fine-textured and very juicy with a sweet, pleasant vinous flavour.

Type: Dessert.
Origin: Dundela, Dundas County, Ontario, Canada, 1796, introduced *c.*1870.
Parentage: ?Fameuse or Saint Lawrence (female) x Unknown.
Flowering: Early.
Other names: M'Intosh, Mac Intosh, Mac Intosh Red, Mac-Intosh, Mac-Intosh Red, MacIntosh, MacIntosh Red, Mackintosh, Mackintosh Red, Makintos, Mc-Intosh Red, McIntosh Red and Mekintos.

McLellan

The flattened fruits, around 7cm (2¾in) across or more, are bright green. The whitish flesh is fairly firm, fine and very tender with a sweet subacid flavour.

Note: Trees are of moderate vigour. The fruits ripen in late autumn.

Type: Dessert.
Origin: Woodstock, CT, USA, *c.*1780.
Parentage: Unknown.
Flowering: Early.
Other names: Lilac, M'Clellan's, M'Lellan, M'Lellan's, Mac-Lellan, MacLellan, Martin, Mc Lellan, McClelan, McClellan and Mek-Lellan.

Megumi

Type: Dessert.
Origin: South Korea, 1967.
Parentage: Ralls Janet x Jonathan.
Flowering: Late.

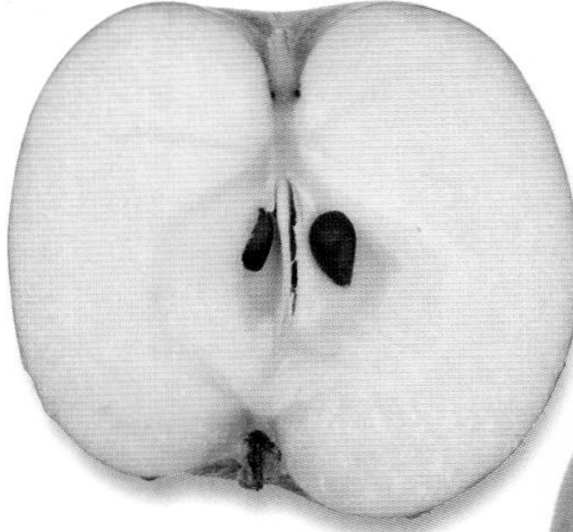

The somewhat conical fruits, around 7cm (2¾in) across, are bright yellowish green with a pinkish red flush. The creamy white flesh is firm and coarse with a very sweet flavour.

Note: Trees crop freely but are weak-growing so should not be on very dwarfing rootstocks.

Melon

Type: Dessert.
Origin: East Bloomfield, Ontario County, NY, USA, *c.*1800, introduced 1845.
Parentage: Unknown.
Flowering: Mid-season.
Other names: Amerikanischer Melonen Apfel, Amerikansk Melonaeble, Melon Apple, Melon de Norton, Melon Norton, Norton, Norton Watermelon, Norton's Melon, Pomme Melon d'Amérique, Pomme Norton and Watermelon.

The rounded but irregular, rather bluntly angular fruits, around 7cm (2¾in) across, are bright lemon yellow tinged with green with a red flush and some flecking and veiny russeting. The yellowish white flesh is firm, fine, crisp and tender with a subacid and aromatic flavour.

Note: Trees are susceptible to frost damage and are weak-growing, so unsuitable for very dwarfing rootstocks. They crop heavily but are prone to biennial bearing; fruits should be thinned. Stored fruits are best before the end of winter.

Mère de Ménage

Type: Culinary.
Origin: Known in the late 1700s.
Parentage: Unknown.
Flowering: Mid-season.
Other names: Bellefleur de France, Brietling, Burton's Beauty, Capp Mammoth, Femme de Ménage, Flanders Pippin, Gelbe Tellerapfel, German Spa, Gloria Mundi, Harlow Pippin, Libra, Livre, Lord Combermere, Menagerie, Mère-de-Ménage, Pfund, Queen Emma, Rambour d'Amérique, Red German, Riesenapfel, Teller and Winter Colmar.

The somewhat irregular, sometimes flattened fruits, up to around 7cm (2¾in) across or more, are bright yellow-green with a strong dark red flush and some streaking and russeting. The greenish flesh is firm, crisp, rather coarse-textured and rather dry with an acid flavour.

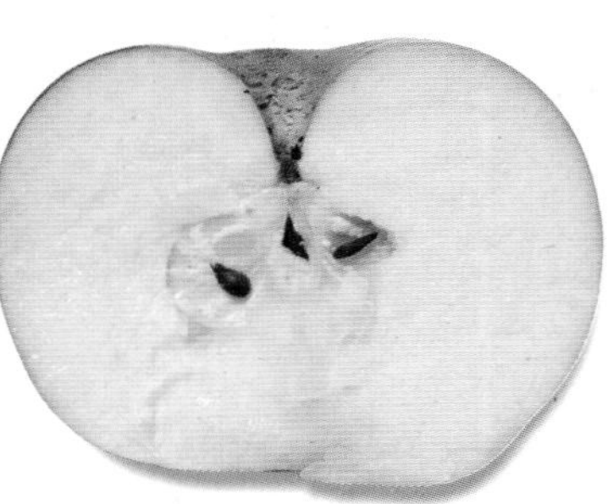

Above: Mère de Ménage is a beautiful apple of first-rate quality.

Note: Fruits can be stored for three to four months.

Merton Beauty

Type: Dessert.
Origin: John Innes Institute, Merton, London, UK, 1933.
Parentage: Ellison's Orange (female) x Cox's Orange Pippin (male).
Flowering: Mid-season.

The rounded fruits, up to around 7cm (2¾in) across, are bright yellow-green with a pinkish red flush and some russeting. The creamy white flesh is firm, fine-textured and juicy with a distinct aniseed flavour.

Note: Trees are moderately vigorous. Fruits are best eaten within two weeks of picking.

Above: The attractive Merton Beauty is easy to grow and always gives a good crop.

Merton Russet

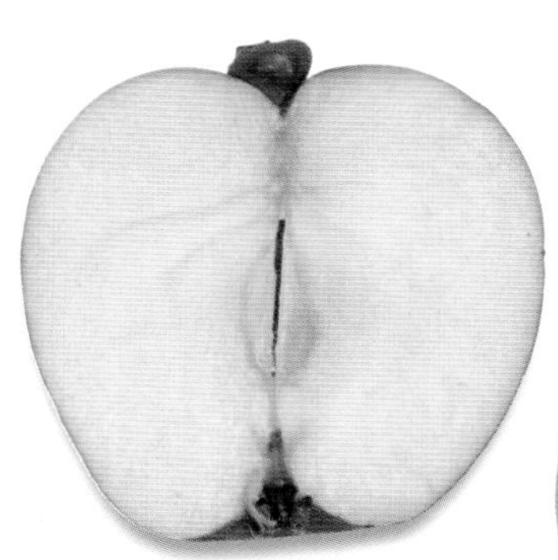

Note: Trees are very vigorous. The variety does best on a dwarfing rootstock.

The rounded, slightly conical fruits, around 6cm (2¼in) across, are bright yellowish green with heavy russeting. The creamy white flesh is firm, crisp and tender with a sweet subacid flavour.

Type: Dessert.
Origin: John Innes Horticultural Institute, Merton, London, UK, 1921 (named 1943).
Parentage: Sturmer Pippin (female) x Cox's Orange Pippin (male).
Flowering: Early.

Mimi

The rounded, slightly irregular fruits, around 6cm (2¼in) across, are bright yellow-green with a strong dark red flush and some flecking. The yellow flesh is fine and soft with a sweet subacid flavour.

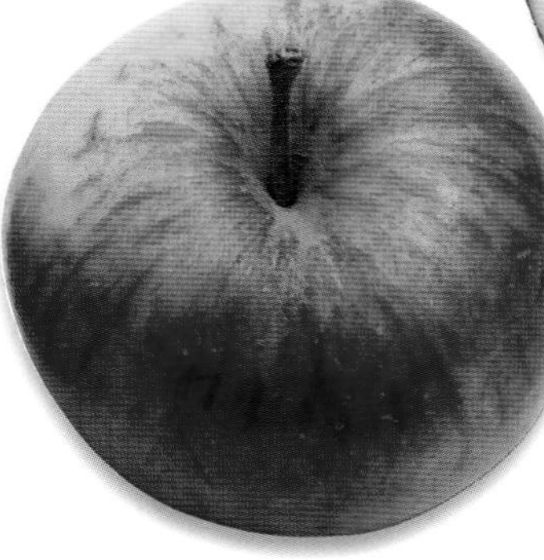

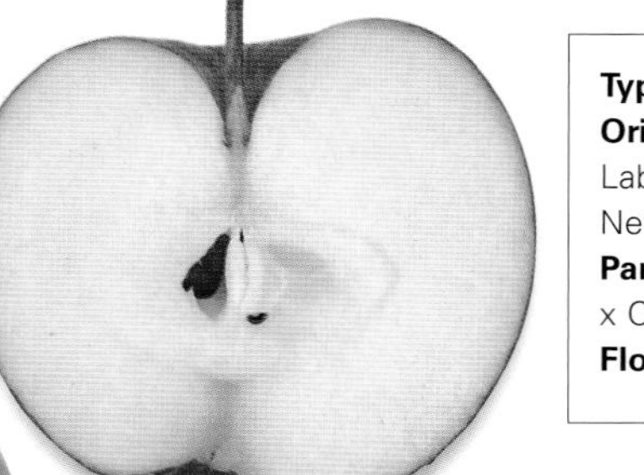

Note: Trees are moderately vigorous.

Type: Dessert.
Origin: Horticultural Laboratory, Wageningen, Netherlands, 1935.
Parentage: Jonathan (female) x Cox's Orange Pippin (male).
Flowering: Mid-season.

Minister von Hammerstein

The flattened, irregular fruits, around 7cm (2¾in) across, are bright yellow-green. The creamy white flesh is firm with a sweet, subacid flavour.

Type: Dessert.
Origin: Geisenheim, Germany, 1882.
Parentage: Landsberger Reinette (female) x Unknown.
Flowering: Early.
Other names: Gammershtein, Hamerstainska, Hammerstein and Hammerstenovo.

Note: Trees are moderately vigorous. They can be vulnerable to mildew and canker in poor growing conditions.

Mitchelson's Seedling

Type: Culinary.
Origin: Kingston upon Thames, Surrey, UK, 1851.
Parentage: Unknown.
Flowering: Mid-season.
Other names: Mitchellson's Seedling, Mitchelson and Mitchelson's Seedling of Hogg.

The rounded but slightly irregular fruits, around 7cm (2¾in) across, are deep yellow-green with a light pinkish red mottled flush and some delicate russeting. The yellowish white flesh is firm and crisp with an acid flavour.

Note: Fully ripe fruits can also be eaten raw. They can be stored for two to three months.

Monarch

Type: Culinary.
Origin: Chelmsford, Essex, UK, 1888, introduced 1918.
Parentage: Peasgood's Nonsuch (female) x Dumelow's Seedling (male).
Flowering: Mid-season.

The rounded fruits, around 7cm (2¾in) across, are bright yellow-green with a red flush and some striping. The white flesh is rather soft, somewhat coarse-textured and juicy with a subacid flavour.

Note: The fruits bruise very easily. They cook to a juicy purée, not as sharp as Bramley. Trees, which are vigorous, are partially self-fertile.

Montfort

Type: Dessert.
Origin: Woodford Green, Essex, UK, *c.*1928.
Parentage: Unknown.
Flowering: Mid-season.

The flattened fruits, around 6cm (2¼in) across, are bright greenish yellow with a red flush. The greenish white flesh is firm and crisp with a subacid flavour.

Note: This variety is self-sterile, so a suitable pollination partner is required for successful fruiting. Fruits can be stored for up to four months.

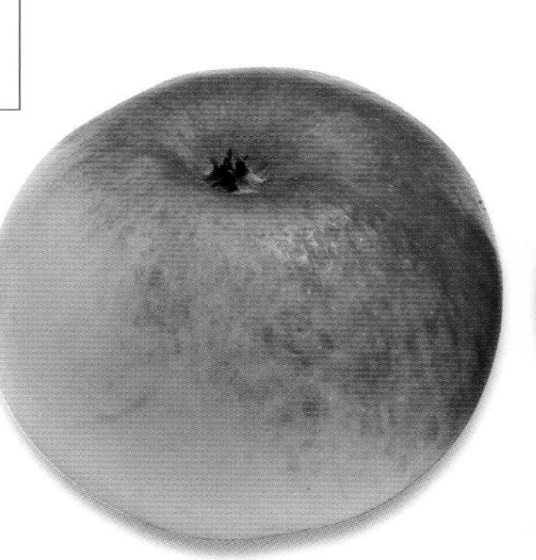

***Above**: Montfort is an apple with an attractive appearance and a mellow flavour.*

Montmedy

The somewhat flattened fruits, around 6cm (2¼in) across, are yellow-green with a strong red flush and some russeting.

Note: Late flowering makes this a suitable variety for growing in areas where spring frosts are likely.

Type: Dessert.
Origin: Italy, recorded 1864.
Parentage: Unknown.
Flowering: Late.

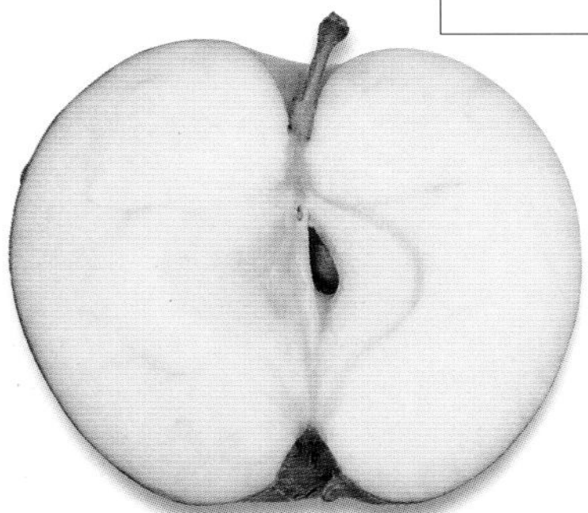

Mors de Veau

The slightly rounded but irregular fruits, around 6cm (2¼in) across, are yellow-green with a strong dark red flush. The greenish white flesh is firm and fine with a subacid flavour.

Note: This variety is rare in cultivation.

Type: Dessert.
Origin: Switzerland, 1948.
Parentage: Unknown.
Flowering: Late.

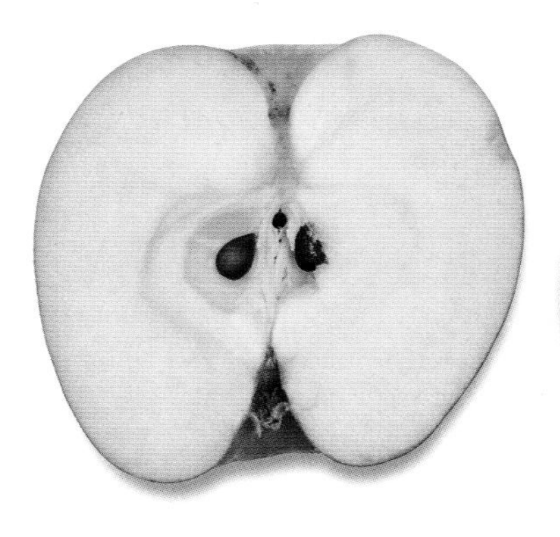

Moss's Seedling

The rounded fruits, around 7cm (2¾in) across, are yellow-green with a red flush that appears as stripes and flecks. The creamy white flesh is crisp and juicy.

Type: Dessert.
Origin: Newport, Shropshire, UK, *c.*1955.
Parentage: Unknown.
Flowering: Mid-season.

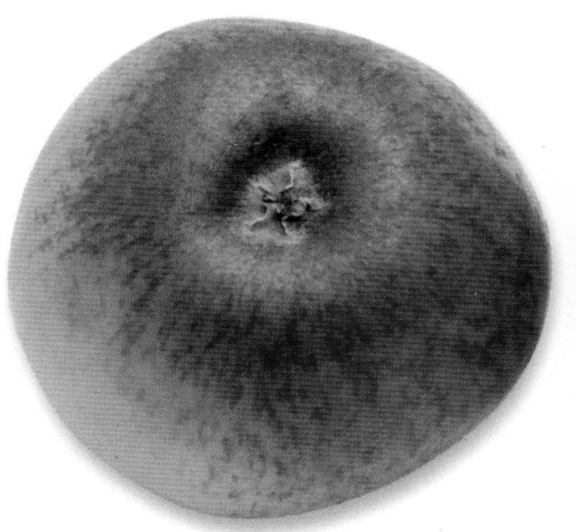
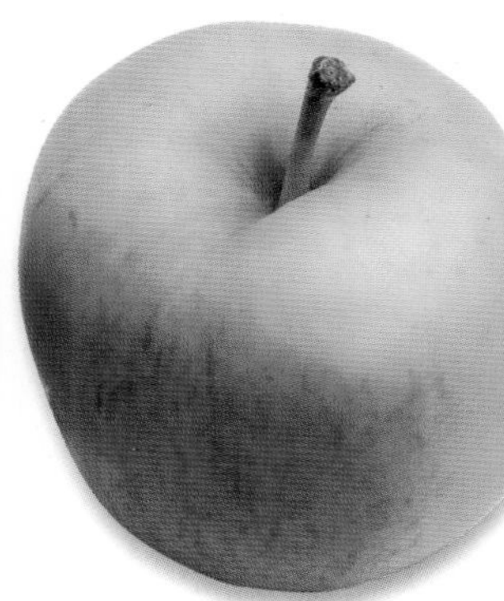
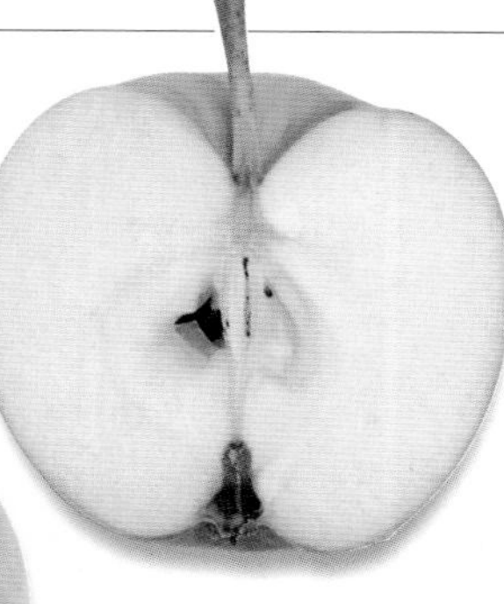

Note: The flavour is similar to that of Cox's Orange Pippin.

Right*: Ripening fruits develop a characteristic striping and flecking on the skins.*

Mother

Type: Dessert.
Origin: Boston, Worcester County, MA, USA, recorded 1844.
Parentage: Unknown.
Flowering: Mid-season.
Other names: American Mother, Gardener's Apple, Mother Apple, Mother of America, Mother of the Americans, Mutter Apfel, Mutterapfel, Queen, Queen Anne and Queen Mary.

The conical, irregular fruits, around 6cm (2¼in) across, are bright yellow-green with a strong dark red flush, sometimes with a mottled or streaked appearance. The yellowish white flesh is fairly firm and very juicy with a distinctive aromatic flavour.

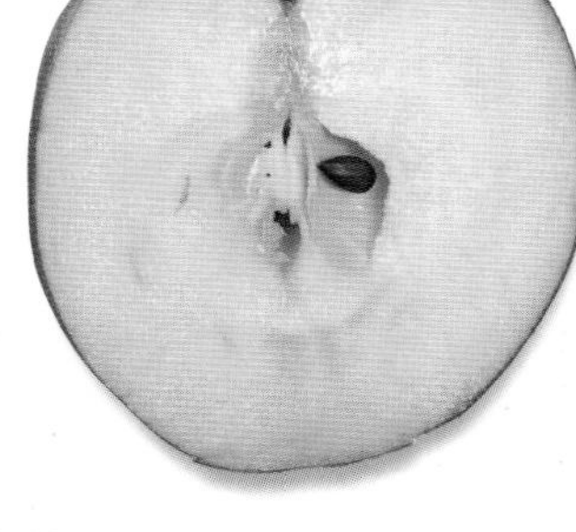

Note: The skins of this variety are very waxy. The fruits need full sun to develop their best flavour. Trees are partially self-fertile.

Moti

Type: Dessert.
Origin: Muntii Apuseni, Romania, 1958.
Parentage: Unknown.
Flowering: Late.
Other names: Moate and Motate.

The rounded, sometimes slightly bumpy fruits, around 7cm (2¾in) across, are bright yellow-green. The creamy white flesh is very firm with a sweet subacid flavour.

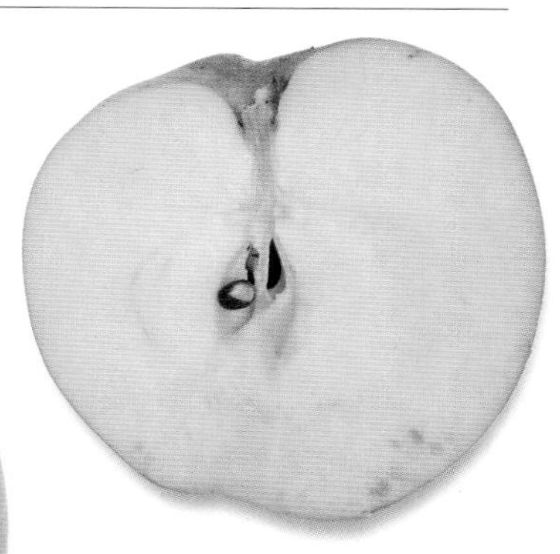
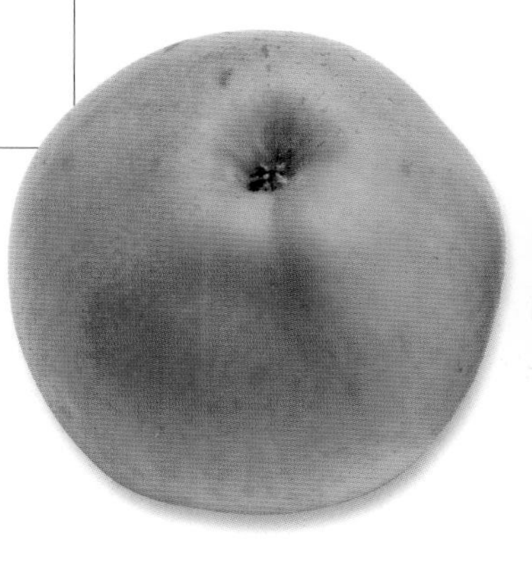

Note: Late flowering makes this a suitable variety for growing in areas where spring frosts are likely.

Mount Rainier

The conical fruits, around 6cm (2¼in) across, are dull yellow-green with a deep red flush. The yellowish flesh is fine and crisp with a very sweet, aromatic flavour.

Type: Dessert.
Origin: ?Netherlands, by 1929.
Parentage: Unknown.
Flowering: Early.

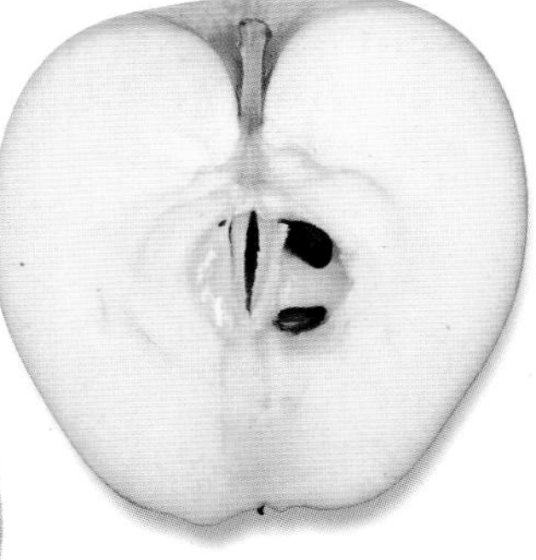

Note: Trees are moderately vigorous.

Right: Fruits have the reddest colouring on the face that is most exposed to the sun.

Museau de Lièvre

Note: Fruits can be stored for five to six months.

The name Museau de Lièvre is generic, referring to a family of apples with a similar shape but differing in colour and flavour. Fruits are tall and conical and about 5cm (2in) across. The creamy white flesh can be very acidic to sweet and fragrant, depending on the variety.

Type: Dessert.
Origin: France.
Parentage: Unknown.
Flowering: Variable.
Selected forms: Museau de Lièvre blanc, Museau de Lièvre jaune, Museau de Lièvre rouge, Museau de Lièvre rouge du Béarn and Museau de Lièvre de septembre.

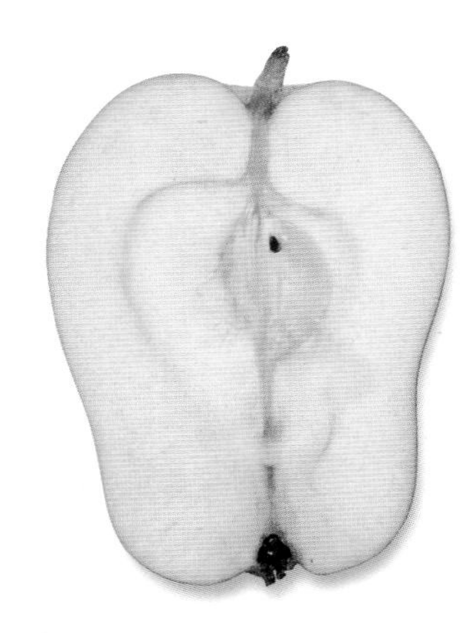

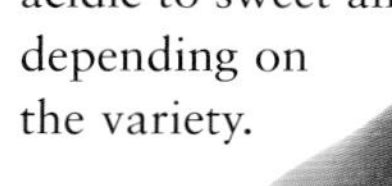

Mutsu

The irregular fruits, up to around 7cm (2¾in) across or more, are bright yellow-green with some russeting. The creamy white flesh is firm, fine-textured and juicy with a slightly sweet, somewhat acid but refreshing and pleasant flavour.

Type: Dessert and culinary.
Origin: Japan, 1930.
Parentage: Golden Delicious (female) x Indo (male).
Flowering: Mid-season.
Other name: Crispin.

Note: This variety is a triploid. It is very susceptible to apple scab, fireblight and powdery mildew. Fruits can be stored for three to six months. They are also suitable for cider making.

Left: This variety is named after the Mutsu Province of Japan, where it is assumed it was first grown.

Neild's Drooper

Type: Culinary and dessert.
Origin: Woburn, Bedfordshire, UK, 1915–16.
Parentage: Unknown.
Flowering: Late.

The somewhat irregular fruits, around 6cm (2¼in) across, are bright yellow-green with a strong red flush and some russeting. The greenish white flesh is crisp, tender and watery with an acid flavour.

Note: The tree has an unusual weeping habit, hence the name.

Left: Ripe fruits of this variety have a very complex flavour.

Nemtesc cu Miezul Rosu

Type: Dessert.
Origin: Romania, 1948 (but probably much older).
Parentage: Unknown.
Flowering: Early.
Other name: Cu Miezul Rosu.

The conical fruits, around 6cm (2¼in) across, are bright yellow-green with a strong red flush and some striping. The flesh is soft and coarse with a subacid flavour.

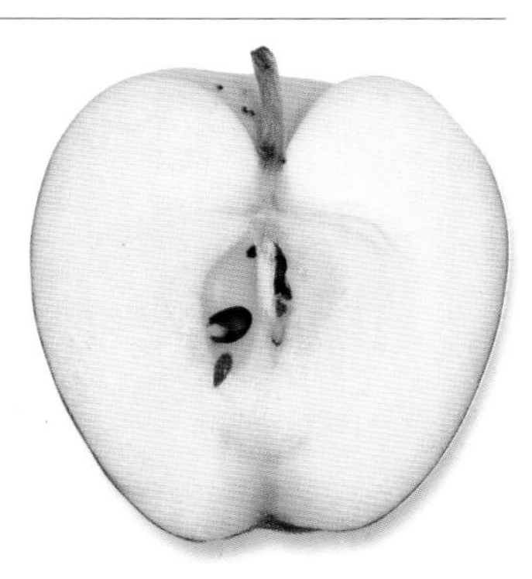

Note: This spur-bearing variety is native to Romania and has been the subject of fertility studies.

Neue Goldparmane

Type: Dessert.
Origin: Germany, 1951.
Parentage: ?Golden Winter Pearmain (female) x Parker's Grey Pippin (male).
Flowering: Early.
Other names: Strauwaldts Goldparmane and Strauwaldts Neue Goldparmane.

The rounded to slightly conical fruits, around 6cm (2¼in) across, are light green with an orange-red flush and some uneven russeting. The pale yellow flesh is firm and fine with a sweet, rich flavour.

Note: Early flowering makes this variety unsuitable for areas where hard spring frosts are likely.

Newton Wonder

The slightly flattened, rounded to irregular fruits, more than 6cm (2¼in) across, are bright green with a strong red, sometimes striped, flush. The creamy white flesh is rather coarse-textured and moderately juicy with a subacid flavour.

Note: The fruits cook well, with a light, fluffy texture, producing a sweeter purée than a Bramley. They store well. Trees, which are vigorous, show good disease resistance. They are partially self-fertile.

Type: Culinary.
Origin: King's Newton, Melbourne, Derbyshire, UK, introduced *c.*1887.
Parentage: ?Dumelow's Seedling (female) x ?Blenheim Orange (male).
Flowering: Mid-season.

Newtown Pippin

The slightly flattened, somewhat irregular fruits, around 6cm (2¼in) across, are bright green with some russeting and darker spotting. The creamy white flesh is firm, fine-textured and juicy with a trace of richness.

Note: This variety was made famous by Thomas Jefferson, who grew them in his orchard at Monticello.

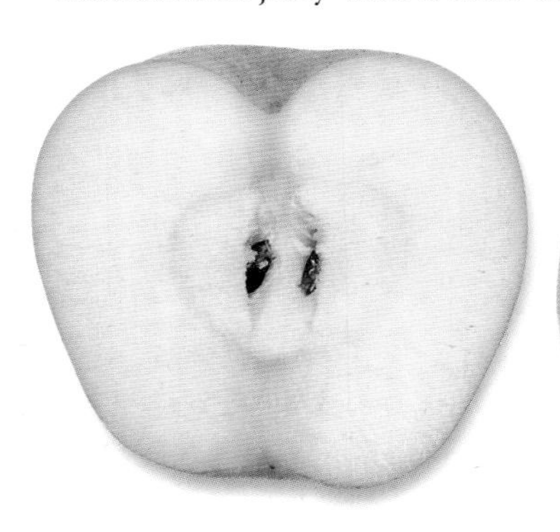

Type: Dessert.
Origin: Newtown, Long Island, New York, USA, by 1759.
Parentage: Unknown.
Flowering: Mid-season.
Other names: Albemarle, American Newtown Pippin, Back Creek, Brookes Pippin, Coxe's Green Newton Pippin, Green Winter Pippin, Hampshire Greening, Hunt's Mountain Pippin, Neujorker Reinette, New York Greening, New York Pippin, Ohio Green Pippin, Pepin Vert de Newtown, Virginia Pippin, White Newtown Pippin and Yellow Newtown Pippin.

Newtown Spitzenburg

The rounded but slightly conical fruits, around 6cm (2¼in) across, are bright green with a deep red flush and some spotting. The creamy yellow flesh is firm and coarse with a sweet subacid flavour.

Note: This variety is sometimes confused with Esopus Spitzenburg. Trees are vigorous but prone to biennial bearing. The fruits store well.

Type: Dessert.
Origin: Newtown, Long Island, NY, USA, recorded 1817.
Parentage: Unknown.
Flowering: Early.
Other names: Barrett's Spitzenberg, Burlington, Burlington Spitzenberg, English Spitzenberg, Joe Berry, Kounty, Kountz, Matchless, Ox-Eye, Queen of the Dessert, Spiced Ox-Eye, Spitzemberg, Spitzenburgh and Vandevere of New York.

Nolan Pippin

Type: Dessert.
Origin: Colchester, Essex, England, UK, 1920.
Parentage: Unknown.
Flowering: Mid-season.

The slightly flattened, somewhat irregular fruits, around 6cm (2¼in) across, are green with uneven cinnamon brown russeting. The white flesh is crisp and hard with a sweet subacid flavour.

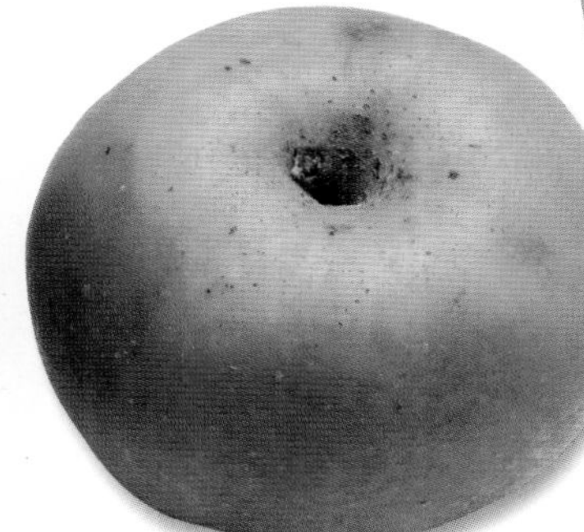

Note: Trees are vigorous. Fruits store well.

Nonpareil

Type: Dessert.
Origin: ?France, introduced into England mid-1500s.
Parentage: Unknown.
Flowering: Early.
Other names: Alter Nonpareil, Bespodobnoe starinnoe, Duc d'Arsell, English Nonpareil, Golden Russet Nonpareil, Groene Reinette, Gruener Reinette, Hunt's Nonpareil, Loveden's Pippin, Nonpareil Old, Original Nonpareil, Pomme Poire, Reinette Franche, Reinette Nonpareil, Reinette Sans Pareille, Reinette Verte and Unvergleichliche Reinett.

The irregular, flattened fruits, up to around 6cm (2¼in) across or more, are yellowish green with uneven russeting and spotting. The greenish white flesh is fine-textured and juicy with a slightly acid and pleasant aromatic flavour.

Note: Trees are healthy and bear well. The fruits store well.

Nonsuch Park

Type: Dessert.
Origin: England, UK, described 1831.
Parentage: Unknown.
Flowering: Mid-season.
Other names: Nonesuch Park and Nonsuch Park Apple.

The slightly flattened fruits, around 6cm (2¼in) across, are bright yellow-green with some spotting. The greenish white flesh is firm and crisp.

Note: This variety commemorates a deer hunting park established by Henry VIII of England. The fruits have waxy skins.

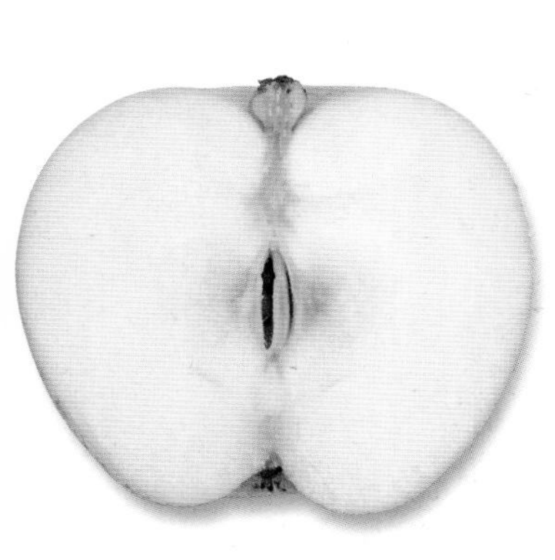

Norfolk Royal

The conical fruits, around 7cm (2¾in) across, are bright yellow-green with a strong red flush that can appear as striping. The creamy white flesh is moderately firm, crisp and very juicy with a sweet and pleasant, melon-like flavour.

Type: Dessert.
Origin: North Walsham, Norfolk, UK, *c.*1908.
Parentage: A chance seedling.
Flowering: Mid-season.

Note: The fruits are best eaten soon after picking as they do not store well. Norfolk Royal Russet is a russeted sport of this variety.

Norman's Pippin

The rounded fruits, around 6cm (2¼in) across, are bright green with light russeting. The greenish white flesh is soft with a sweet, rich flavour.

Type: Dessert.
Origin: ?Flanders, recorded 1900.
Parentage: Unknown.
Flowering: Mid-season.

Note: Trees can lack vigour. Fruits can be used for juicing or cider making.

Right: Fruits show even colouring and a waxy skin when ripe.

Normandie

The slightly irregular fruits, around 6cm (2¼in) across, are bright green with a strong dark red flush and some striping. The greenish white flesh is soft and dry with an insipid flavour.

Note: Late flowering makes this a suitable variety for growing in areas with late frosts.

Type: Dessert.
Origin: ?France, 1948.
Parentage: Unknown.
Flowering: Very late.

Northern Greening

Type: Culinary.
Origin: ?England, UK, first recorded 1826.
Parentage: Unknown.
Flowering: Mid-season.
Other names: Cowarn Queening, Cowarne Queening, Cowarne Seedling, Gruener Englischer Pepping, John, Kirk Langley Pippin, Langley Pippin, Old Northern Greening, Verte du Nord, Walmer Court and Woodcock.

The rounded, slightly irregular fruits, around 6cm (2¼in) across, are yellow-green with a red flush and some spotting and russeting. The creamy white flesh is moderately firm, a little coarse-textured and juicy with an acid flavour.

Note: This variety is probably the ancestor of many Victorian culinary apples. The fruits store well. Trees are productive and disease resistant. They tolerate cold winters.

Northern Spy

Type: Dessert.
Origin: East Bloomfield, NY, USA, *c.*1800, introduced 1840.
Parentage: Unknown.
Flowering: Late.
Other names: King Apple, King's Apple, Severnui Razvedchik, Severnui Shpion, Spaeher des Nordens, Spaher des Nordens and Spy.

Note: This variety is resistant to woolly aphid and has been used as a parent in the breeding of resistant rootstocks and varieties. Trees, which are vigorous, are prone to biennial bearing. Fruits are late to ripen and store well.

The irregular fruits, around 7cm (2¾in) across, are yellow-green with a strong dark red flush and some spotting. The yellow flesh is fairly firm, juicy and sweet with a pleasant flavour.

Above: Fruits develop a red flush where they are exposed to the sun.

Nottingham Pippin

Type: Dessert.
Origin: ?Nottingham, UK, by 1815.
Parentage: Unknown.
Flowering: Mid-season.
Other names: Pepping von Nottingham and Pippin of Nottingham.

The irregular, somewhat flattened fruits, around 6cm (2¼in) across, are bright yellow-green with some russeting. The white flesh is fine and tender with a sweet, vinous flavour.

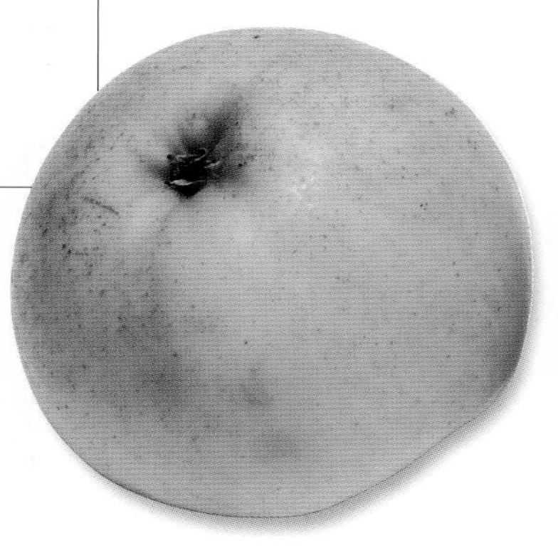

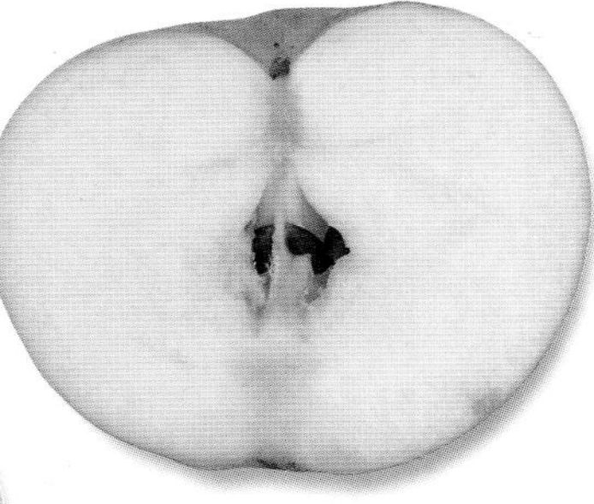

Note: Trees are vigorous and bear freely. This variety is self-sterile, so a suitable pollination partner is required.

Nova Easygro

Type: Dessert.
Origin: Canadian Department Agricultural Research Station, Kentville, Nova Scotia, 1956, introduced 1971.
Parentage: Spartan (female) x Scab Resistant seedling (male).
Flowering: Mid-season.

The slightly irregular fruits, around 6cm (2¼in) across, are bright greenish yellow with a strong red flush and some striping. The whitish flesh is firm, crisp and moderately juicy with a subacid and pleasant flavour.

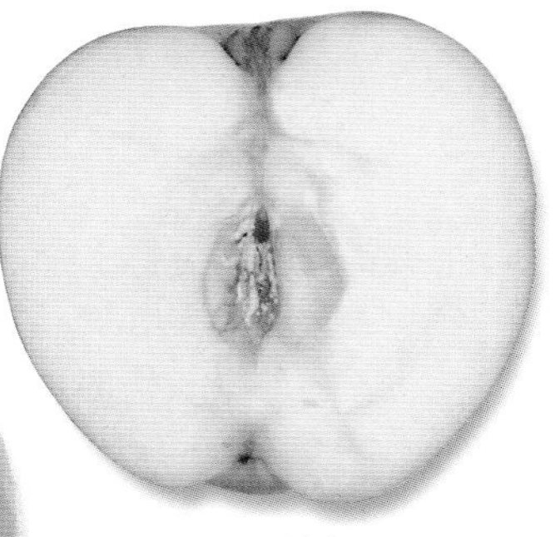

Note: Trees, which are vigorous, are resistant to scab and cedar apple rust, and show some resistance to fireblight and mildew.

Old Fred

The flattened fruits, up to 7cm (2¾in) across, are dull yellow-green with a pinkish red flush and some striping. The creamy white flesh is firm and fine with a subacid flavour.

Note: Old Fred, rare in cultivation, was named after the nurseryman who bred it.

Type: Dessert.
Origin: Eynsham, Oxford, UK, exhibited 1944.
Parentage: Allington Pippin (female) x Court Pendu Plat (male).
Flowering: Mid-season.

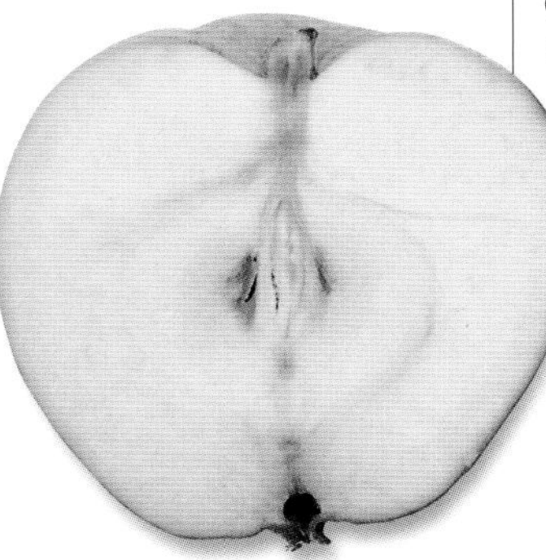

Ontario

The somewhat flattened, irregular fruits, around 7cm (2¾in) across or more, are bright yellow-green with a very strong dark red flush and some striping. The creamy white flesh is crisp and very juicy with a rather acid flavour.

Type: Dessert and culinary.
Origin: Paris, Ontario, Canada, *c.*1820.
Parentage: Wagener (female) x Northern Spy (male).
Flowering: Mid-season.
Other name: Ontarioapfel.

Note: The flowers are weather resistant. The fruits, which store well, tend to bruise easily. They cook well, breaking up completely. Trees are moderately vigorous. There is also a tetraploid clone (with four sets of chromosomes).

Opalescent

Type: Dessert.
Origin: Xenia, OH, USA, introduced 1899.
Parentage: Unknown.
Flowering: Mid-season.

The conical fruits, around 6cm (2¼in) across, are bright yellow-green with a strong red flush. The creamy white flesh is firm and crisp with a sweet, strawberry-like taste.

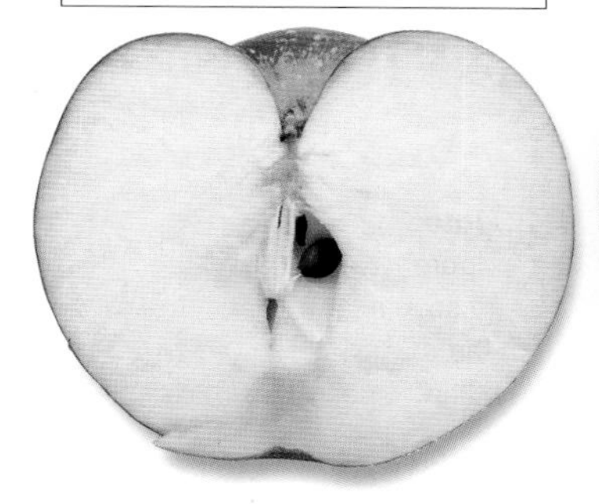

Note: This variety is a parent of Shenandoah. The fruits are also suitable for cooking.

Right: Its excellent flavour has made Opalescent a very popular variety.

Orin

Type: Dessert.
Origin: Fukushima Prefecture, Japan, before 1942.
Parentage: Golden Delicious (female) x Indo (male).
Flowering: Very early.

The conical fruits, around 6cm (2¼in) across, are bright yellow-green with russeting and spotting. The creamy white flesh is firm and juicy with a sweet and aromatic, honeyed flavour.

Note: This variety is a triploid, so two pollinators will be needed for fruiting. Trees are tip-bearing.

Orleans Reinette

Above: This variety develops a complex flavour as it ripens.

The somewhat flattened fruits, around 6cm (2¼in) across, are yellow-green with an orange-red flush and some striping and light russeting. The creamy white flesh is firm, fine-textured and only a little juicy with a pleasant, Blenheim-like flavour.

Note: Trees are moderately vigorous, hardy and crop well. They fruit best in a warm location.

Type: Dessert.
Origin: Unknown, but assumed to be French, first described 1776.
Parentage: Unknown.
Flowering: Late.
Other names: Aurore, Cardinal Pippin, Doerell's Grosse Gold Reinette, Madam Calpin, New York Reinette, Pearmain d'Or, Reinette d'Aix, Reinette de Breil, Reinette Golden, Reinette Triomphante, Ronde Belle-Fleur, Starklow's Bester, Triumph Reinette, Winter Ribston, Yellow German Reinette and Zimnii Shafran.

Osnabrücker Reinette

The rounded fruits, around 6cm (2¼in) across, are bright green with some uneven russeting. The creamy white flesh is fine and tender with a sweet and very vinous flavour.

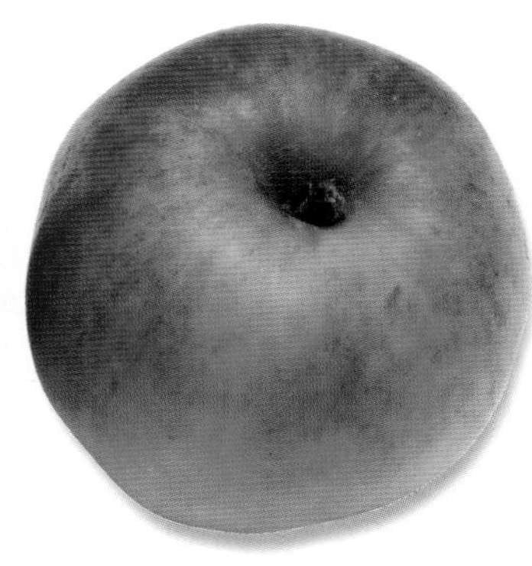

Note: Fruits can be stored for up to five months. Trees are vigorous initially, developing a large, rounded crown.

Type: Dessert.
Origin: Osnabrück, Hannover, Germany, by 1802.
Parentage: Chance seedling.
Flowering: Mid-season.
Other names: De Grawe Foos-Renet, Franz Joseph von Eggers Reinette, Französische Goldereinette, Gold-Reinette, Graawe Fos-Renet, Graue Osnabrücker Reinette, Graue Reinette von Canada, Osnabrücker grau überzogene Reinette, Reinette Aigre, Reinette d'Osnabrück, Reinette Grise d'Osnabrück, Renet Osnabrukskii and Rotgraue Kelch-Reinette.

Oxford Conquest

The slightly irregular, rather flattened fruits, around 6cm (2¼in) across or more, are dull yellow-green with an orange-red flush and some russeting. The creamy white flesh is rather tough with an acid flavour.

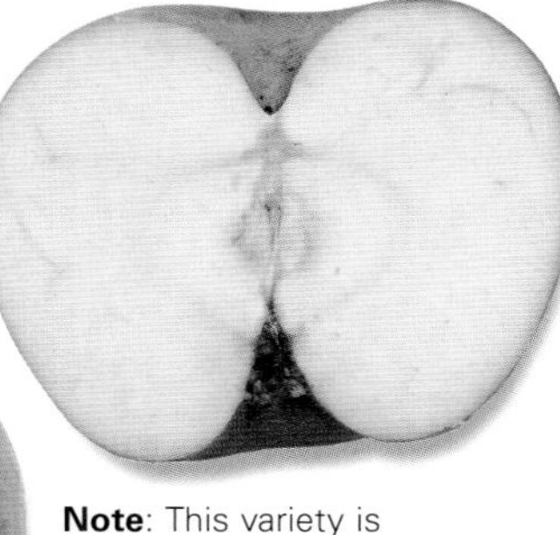

Note: This variety is self-sterile, so a suitable pollination partner is needed for reliable fruiting.

Type: Dessert.
Origin: Eynsham, Oxford, UK, 1927.
Parentage: Blenheim Orange (female) x Court Pendu Plat (male).
Flowering: Mid-season.

Oxford Hoard

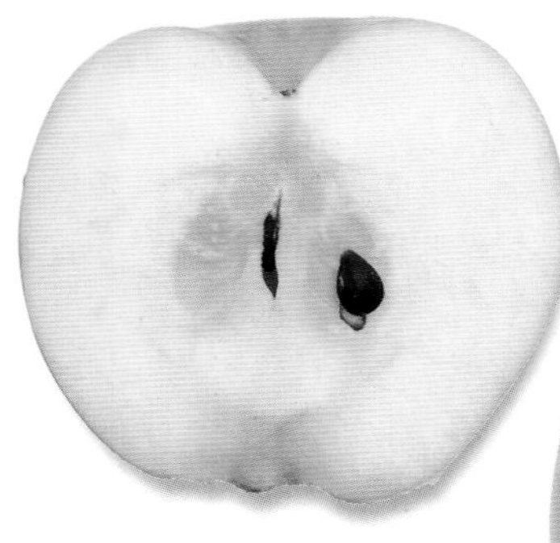

The irregular fruits, around 6cm (2¼in) across, are bright yellowish green with some spotting. The creamy white flesh is coarse, firm and tough with a sweet and aromatic flavour.

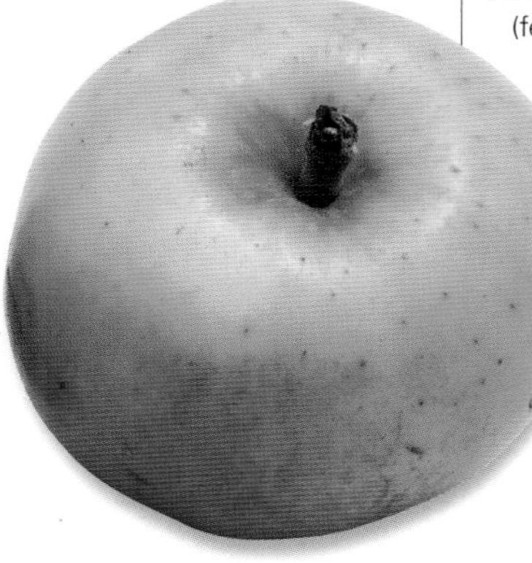

Note: Trees are very vigorous. The skins of the fruits can be quite waxy.

Type: Dessert.
Origin: Eynsham, Oxford, UK, exhibited 1943.
Parentage: Sturmer Pippin (female) x Golden Russet (male).
Flowering: Early.

Parker's Pippin

Type: Dessert.
Origin: ?England, before 1800.
Parentage: Unknown.
Flowering: Early.
Other names: Broker's Pippin, Graue Reinette, Jadrnac Parkeruv, Kozhanyi renet, Lederapfel, Parker, Parker Peppin, Parker's Grey Pippin, Parkerova, Parkers Grauer Pepping, Parsker's Pippin, Pepin Gris de Parker, Peppina Parker, Pippin de Parker, Pomme Parker, Poppina Parker, Reinette Grise de Pfaffenhofen, Sanct-Nicolas Reinette, Spencer's Pippin and Zizzen-Apfel.

The rounded to oblong fruits, around 6cm (2¼in) across, are light green with russeting. The creamy white flesh is fairly crisp and firm with a subacid and slightly aromatic flavour.

Note: This variety performs well on a dwarfing rootstock. It is one of the parents of Neue Goldparmane.

Peck's Pleasant

Type: Dessert.
Origin: Rhode Island, USA, recorded 1832.
Parentage: Unknown.
Flowering: Mid-season.
Other names: Dutch Greening, Peck, Peck Pleasant, Waltz Apple and Watts Apple.

The rounded, slightly irregular fruits, around 6cm (2¼in) across, are bright green to yellow with some lighter flecking and spotting; an orange-red flush can sometimes partly deepen to pink. The yellowish flesh is crisp with a sweet, subacid, aromatic flavour.

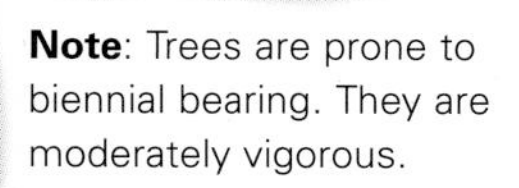

Note: Trees are prone to biennial bearing. They are moderately vigorous.

Pero Dourado

Type: Dessert.
Origin: Portugal, 1952.
Parentage: Unknown.
Flowering: Mid-season.

Above: The pointed shape of the fruits is characteristic.

The conical to oblong fruits, around 6cm (2¼in) across, are bright yellow-green with a pinkish red flush and some russeting. The creamy white flesh is firm and fine with a subacid flavour.

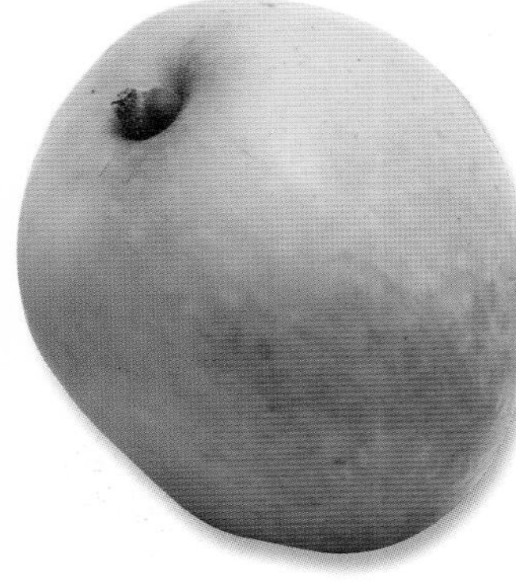

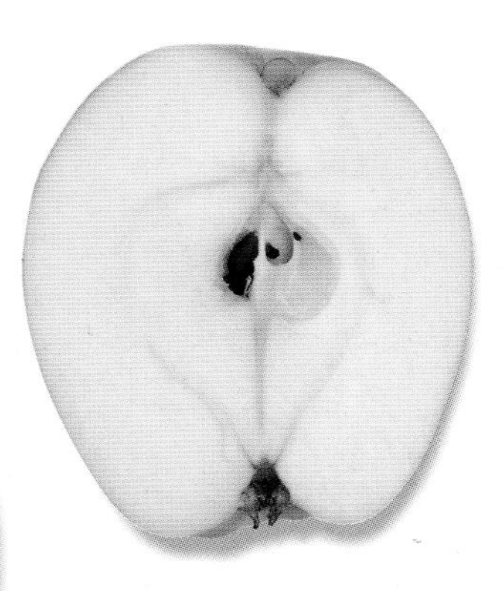

Note: This variety is rare in cultivation but has been used as a subject for research into cultivar identification.

Petit Pippin

The rounded to slightly irregular fruits, around 6cm (2¼in) across, are bright yellow-green with some darker spotting and russeting. The creamy white flesh is soft with a sweet, aromatic flavour.

Note: The fruits are best in late summer.

Type: Dessert.
Origin: East Malling Research Station, Kent, UK, 1948.
Parentage: Unknown.
Flowering: Mid-season.

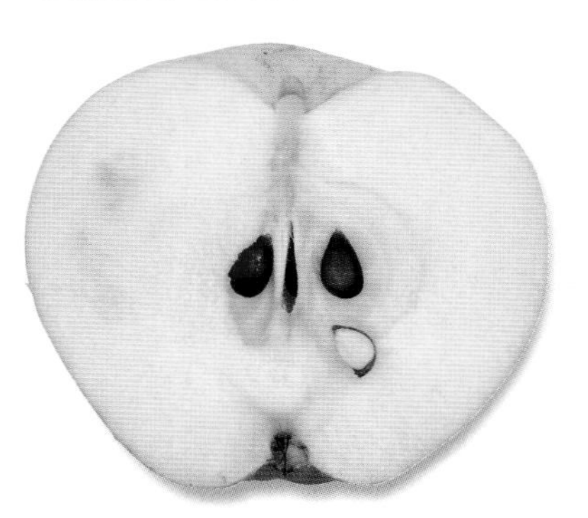

Pig's Nose Pippin

The roughly conical fruits, around 6cm (2¼in) across, are yellow-green with a bright red flush and some russeting in dots and patches. The creamy white flesh is fine and crisp with a sweet flavour.

Type: Dessert.
Origin: ?Hereford, UK, described 1884.
Parentage: Unknown.
Flowering: Mid-season.

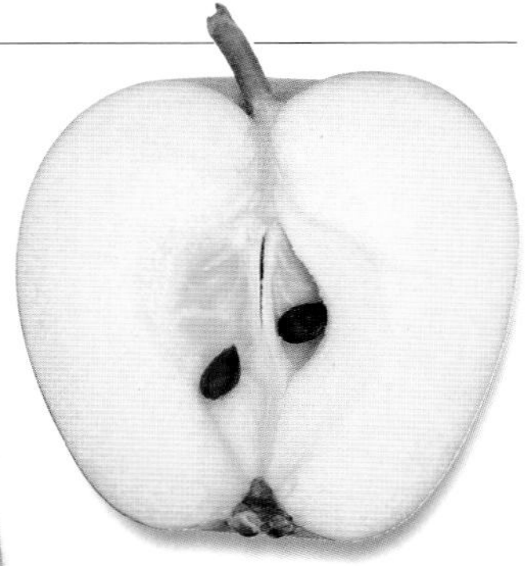

Note: Fruits can be stored until mid-winter only.

Right: Pig's Nose Pippin is often grown simply for the appeal of its curious name.

Pigeon de Jérusalem

Type: Dessert.
Origin: France, recorded late 1600s.
Parentage: Unknown.
Flowering: Mid-season.
Other names: Coeur de Pigeon, Gros Pigeon Rouge, Gros Pigeonnet rouge, Gros-Coeur de Pigeon, Jérusalem, Pigeon, Pigeon d'Hiver, Pigeon Rouge, Pigeonnet de Jérusalem and Pomme de Jérusalem.

The roughly conical, very irregular fruits, around 6cm (2¼in) across, are light green with a strong pinkish red flush and some striping; the whole surface is covered with a bluish bloom. The white flesh is firm and fine with a subacid and slightly perfumed flavour.

Note: Trees are vigorous and crop freely. The fruits are also suitable for cooking.

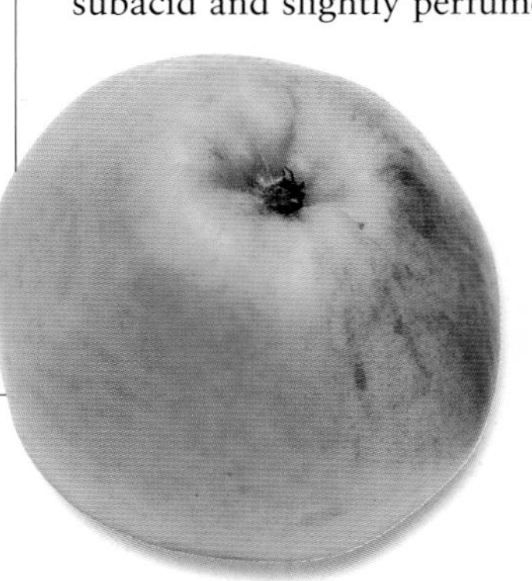

Pink Lady

Type: Dessert.
Origin: Australia, 1979.
Parentage: A selection of Cripps Pink (Lady Williams x Golden Delicious).
Flowering: Mid-season.

The slightly conical to oblong fruits, around 6cm (2¼in) across, are yellow with a distinctive pinkish orange flush. The creamy white flesh is crisp, firm and very sweet.

Note: This variety needs a long summer to ripen and produce the characteristic flush. It is susceptible to rosy apple aphid and scab. Trees crop heavily and the fruits store well. It is an important commercial variety.

Pinner Seedling

Type: Dessert.
Origin: Pinner, Middlesex, UK, 1810.
Parentage: Unknown.
Flowering: Mid-season.
Other names: Carel's Seedling, Carle's Seedling, Carrel's Seedling, Carrell's Seedling, Pinner and Pinners Seedling.

The rounded, slightly conical, sometimes irregular fruits, around 7cm (2¾in) across, are bright green with a pinkish red flush and some russeting (sometimes almost covering the fruit). The whitish flesh, tinged with green, is crisp, juicy and sweet in flavour.

Note: The fruits store well. They can be used for juicing and cider making.

Pitmaston Pineapple

Type: Dessert.
Origin: Witley, UK, *c.*1785.
Parentage: Golden Pippin (female) x Unknown.
Flowering: Early.
Other names: Ananas de Pitmaston, Pine-Apple, Pineapple, Pineapple Pippin, Pitmaston Pine, Pitmaston Pine Apple and Reinette d'Ananas.

The somewhat conical fruits, around 6cm (2¼in) across, are dull yellow-green with some fine russeting. The creamy white to yellowish flesh is firm and juicy with a sweet, rich, distinctive flavour.

Note: The flavour is similar to that of a pineapple, hence the name. Trees are small and upright, so suitable for small gardens.

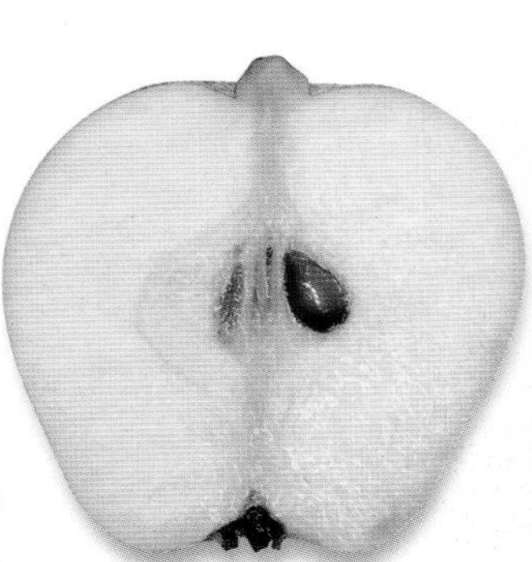

Far right: This old English variety has an unusual flavour.

Pixie

The somewhat flattened fruits, around 6cm (2¼in) across, are bright greenish yellow with a strong red flush and some striping and russeting. The creamy white flesh is crisp and fairly juicy with a good aromatic, somewhat sharp flavour.

Type: Dessert.
Origin: National Fruit Trials, Wisley, Surrey, UK, 1947.
Parentage: Cox's Orange Pippin or Sunset (female) x Unknown.
Flowering: Mid-season.

Note: This variety, which grows well on a dwarfing rootstock, is disease resistant. Trees can overcrop, so fruit should be thinned. The fruits store well.

Pomme Crotte

The flattened, irregular fruits, up to around 7cm (2¾in) across or more, are bright yellow-green with a strong dark red flush. The creamy white flesh is sweet and juicy.

Type: Dessert.
Origin: France, 1947.
Parentage: Unknown.
Flowering: Mid-season.

Note: Trees are moderately vigorous.

Right: Fruits turn dark red where the skin is exposed to the sun.

Pomme d'Amour

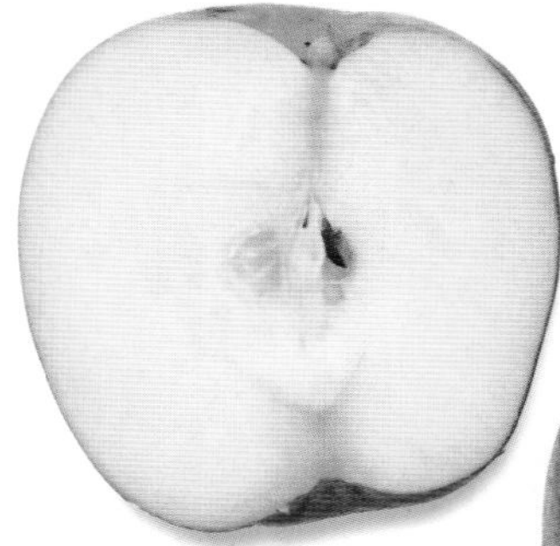

The rounded, slightly irregular fruits, around 6cm (2¼in) across, are bright yellow-green with an orange-red flush and some striping. The yellowish white flesh is firm with a slightly sweet, slightly subacid flavour.

Type: Dessert.
Origin: France, 1947.
Parentage: Unknown.
Flowering: Mid-season.

Note: Trees are moderately vigorous.

Far right: The name of this variety, Pomme d'Amour, means 'toffee apple' in French.

Pomme d'Enfer

Type: Dessert.
Origin: Allier, France, 1948.
Parentage: Unknown.
Flowering: Mid-season.

The rounded to slightly conical fruits, around 7cm (2¾in) across or more, are dull green with a strong red flush and mottled russeting. The greenish white flesh has an acid flavour.

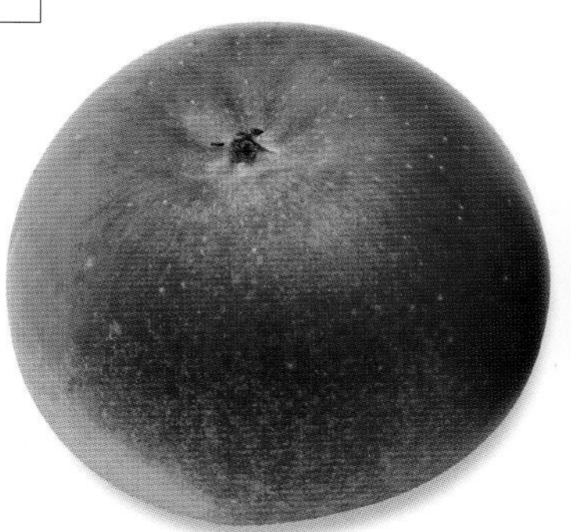

Note: Trees can lack vigour. Picked fruits have been used in France to scent the interiors of wardrobes.

Pomme de Choux à Nez Creux

Type: Culinary.
Origin: Cher, France, 1948.
Parentage: Unknown.
Flowering: Late.

The flattened, rather irregular fruits, around 7cm (2¾in) across or more, are bright yellow-green with some russeting. The yellowish flesh is soft, juicy and acid.

Note: Late flowering makes this variety suitable for growing where spring frosts are likely.

Left: Fruits of this variety show variegated colouring.

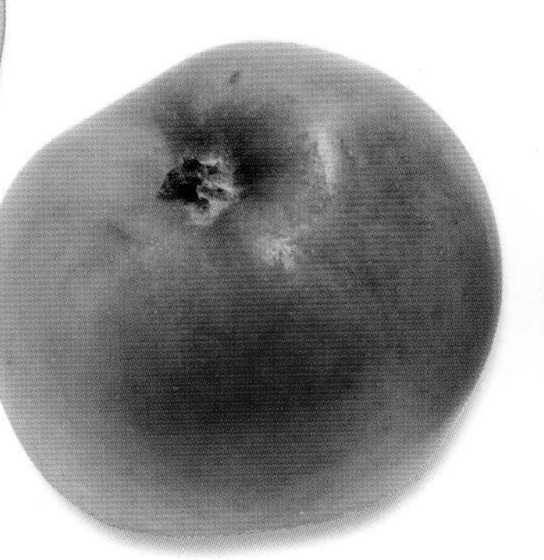

Pomme de Fer

Type: Dessert.
Origin: France, described 1948.
Parentage: Unknown.
Flowering: Mid-season.

The flattened fruits, around 7cm (2¾in) across, are bright yellow-green with a red flush and some striping. The flesh is creamy white.

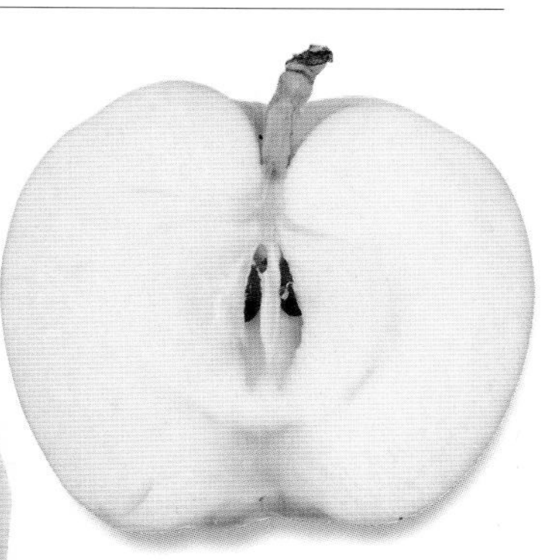

Note: This variety is a parent of Canada Baldwin. The fruits store well and can also be used in cooking.

Pomme de Feu

The rounded fruits, around 6cm (2¼in) across, are bright yellowy green with waxy skins. The creamy white flesh is firm and fine, with a slightly subacid flavour.

Note: This variety is rare in cultivation.

Type: Culinary.
Origin: France, 1948.
Parentage: Unknown.
Flowering: Late.

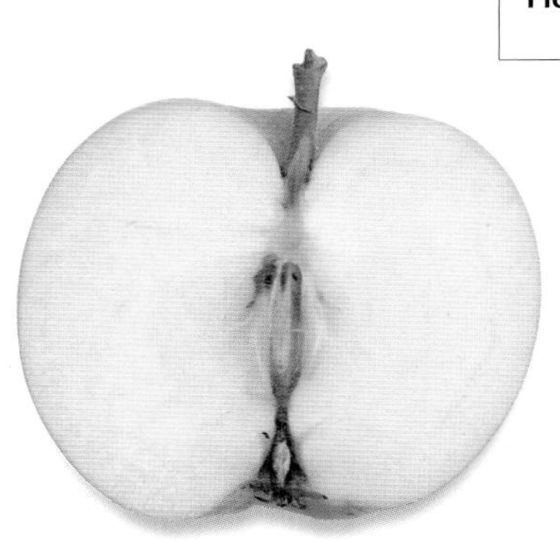

Pomme de Glace

The slightly flattened, slightly irregular fruits, around 7cm (2¾in) across, are bright green with some russeting. The greenish white flesh is firm with an acid flavour.

Type: Culinary.
Origin: France, 1948.
Parentage: Unknown.
Flowering: Late.

Note: This variety gets its name (which translates as 'ice apple') from the frosty appearance of the fruits. These can be stored for three to four months.

Pomme Noire

The rounded, slightly irregular fruits, around 6cm (2¼in) across, are bright green with a very dark red flush. The pale green flesh is grainy with a sweet flavour.

Type: Dessert.
Origin: France, 1973.
Parentage: Unknown.
Flowering: Late.
Other names: Grosse Pomme Noire, Black American, Black Apple and Violette Glacée.

Note: Fruits can be stored for up to five months. Trees are moderately vigorous.

Pommerscher Krummstiel

Type: Dessert.
Origin: Berlin Technical University, Germany, 1967.
Parentage: Unknown.
Flowering: Mid-season.

The irregular fruits, around 7cm (2¾in) across or more, are bright green with a strong red flush. The whitish flesh has a fairly rich, slightly aromatic flavour.

Note: Fruits store well. The flavour is best four weeks after picking.

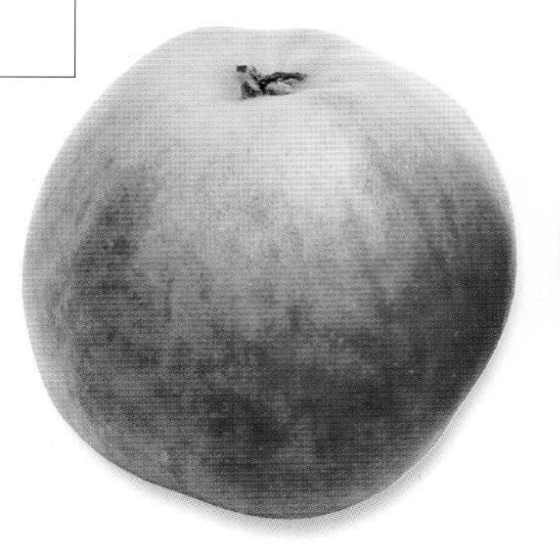

Prairie Spy

Type: Dessert.
Origin: University of Minnesota Fruit Breeding Farm, Excelsior, USA, 1951.
Parentage: Unknown.
Flowering: Mid-season.

The rounded fruits, around 6cm (2¼in) across, are bright yellow-green with a strong dark red flush and some russeting. The greenish white flesh is firm and fine with a sweet, subacid flavour.

Note: Trees are vigorous and long-lived; they bear young. The flavour of the fruits improves in storage.

Present van Holland

Type: Dessert.
Origin: Netherlands, 1945.
Parentage: Present van Engeland (female) x Brabant Bellefleur (male).
Flowering: Early.

The conical to oval fruits, around 6cm (2¼in) across, are bright yellow-green with a bright red flush. The white flesh is firm and coarse with a subacid flavour.

Note: Trees can lack vigour, so this variety is unsuitable for growing on a dwarfing rootstock.

Right: The oval shape of the fruits is characteristic.

Prinses Beatrix

The uneven fruits, up to around 7cm (2¾in) across or more, are yellow-green with a bright red flush and some striping and flecking. The creamy white flesh is firm, fine and rather tough with a subacid flavour.

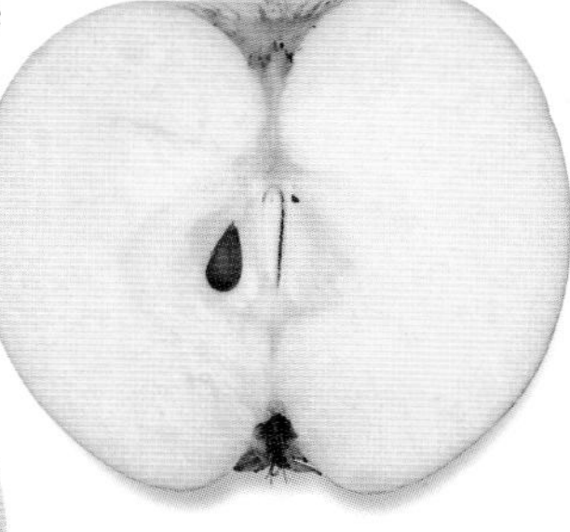

Type: Dessert.
Origin: Horticultural Laboratory, Wageningen, Netherlands, 1935.
Parentage: Cox's Orange Pippin (female) x Jonathan (male).
Flowering: Mid-season.

Note: This attractive variety is rather unusual in cultivation.

Prinses Irene

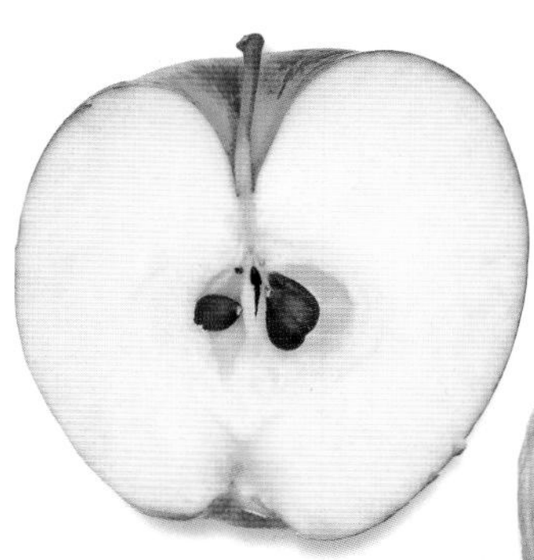

Note: Trees are weak-growing, so are unsuitable for use with a dwarfing rootstock.

The slightly irregular fruits, around 6cm (2¼in) across, are bright yellow-green with a red flush and some striping and flecking. The greenish white flesh is crisp with a subacid flavour.

Type: Dessert.
Origin: Horticultural Laboratory, Wageningen, Netherlands, 1935.
Parentage: Jonathan (female) x Cox's Orange Pippin (male).
Flowering: Early.

Prinses Margriet

The rounded to slightly flattened fruits, around 6cm (2¼in) across, are yellow-green with a red flush and some flecking and russeting. The cream flesh is tinged green and has an acid flavour.

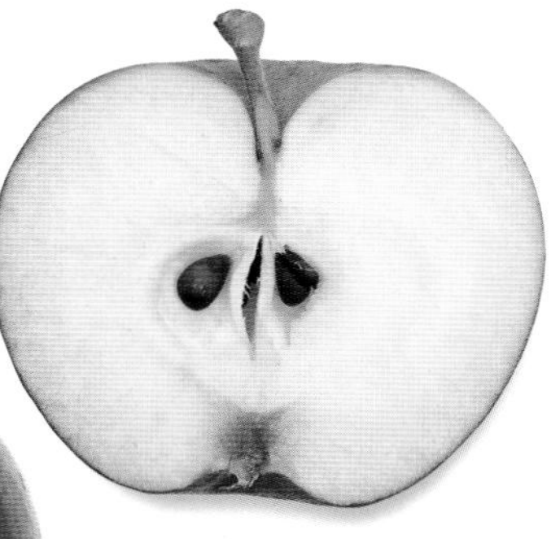

Type: Dessert.
Origin: Horticultural Laboratory, Wageningen, Netherlands, 1935, introduced 1955.
Parentage: Jonathan (female) x Cox's Orange Pippin (male).
Flowering: Mid-season.

Note: Trees are moderately vigorous.

Prinses Marijke

Type: Dessert.
Origin: Horticultural Laboratory, Wageningen, Netherlands, 1935, introduced 1952.
Parentage: Jonathan (female) x Cox's Orange Pippin (male).
Flowering: Mid-season.

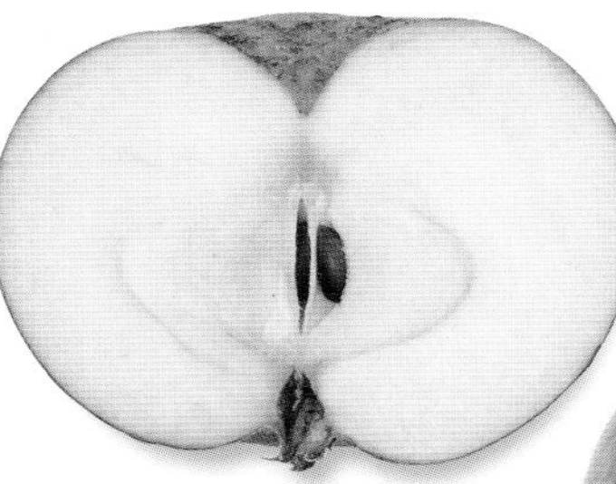

The rounded, slightly flattened fruits, around 7cm (2¾in) across, are bright green with a dark red flush and some russeting. The creamy white flesh is fine, firm and crisp with a fairly sweet, subacid, aromatic flavour.

Note: Trees are weak-growing, so are unsuitable for use with a dwarfing rootstock.

Queen Cox

Type: Dessert.
Origin: Appleby Fruit Farm, Kingston Bagpuize, Berkshire, UK, 1953.
Parentage: ?Ribston Pippin (female) x Unknown.
Flowering: Mid-season.

The rounded fruits, around 6cm (2¼in) across, are green with a red flush and some striping. The creamy white flesh is firm, slightly acid and juicy with a rich, aromatic flavour.

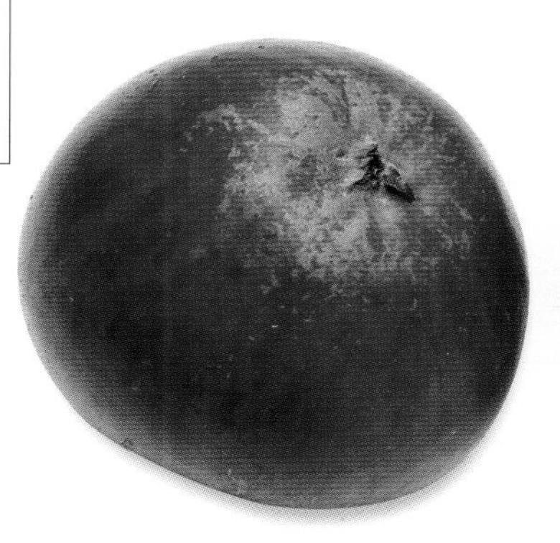

Note: This variety is a more highly coloured clone of Cox's Orange Pippin.

Racine Blanche

Type: Culinary.
Origin: France, 1947.
Parentage: Unknown.
Flowering: Late.

The slightly flattened, somewhat irregular fruits are bright yellow-green with a red flush and some russeting. The yellowish flesh is rather coarse with an acid flavour.

Note: Late flowering makes this variety suitable for growing in areas where late frosts are likely.

Rampale

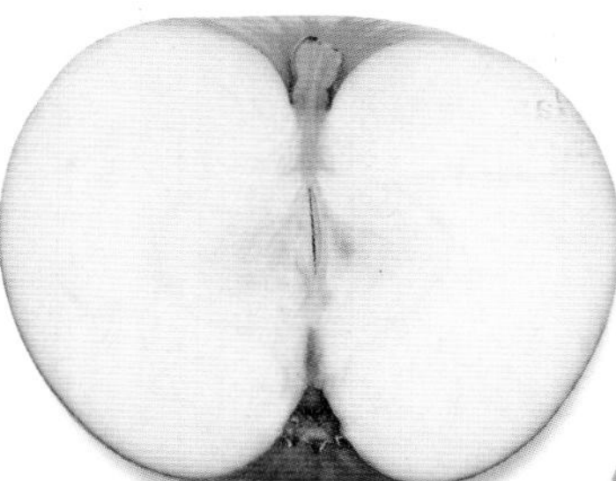

The flattened, somewhat irregular fruits, up to around 6cm (2¼in) across, are yellow-green with a red flush and some flecking and russeting. The whitish flesh is crisp with a sweet, vinous, perfumed flavour.

Type: Dessert.
Origin: France, 1947.
Parentage: Unknown.
Flowering: Late.

Note: Late flowering makes this variety suitable for growing in areas where late frosts are likely.

Reale d'Entraygues

The conical to oblong fruits, up to around 8cm (3in) across or more, are bright yellow-green with some spotting. The yellowish white flesh is very firm with a very sweet, perfumed flavour.

Note: This variety is normally grown as a standard. Avoid hot sites.

Type: Dessert.
Origin: France, recorded 1947 but probably much older.
Parentage: Unknown.
Flowering: Mid-season.
Other name: Reinette de Pons.

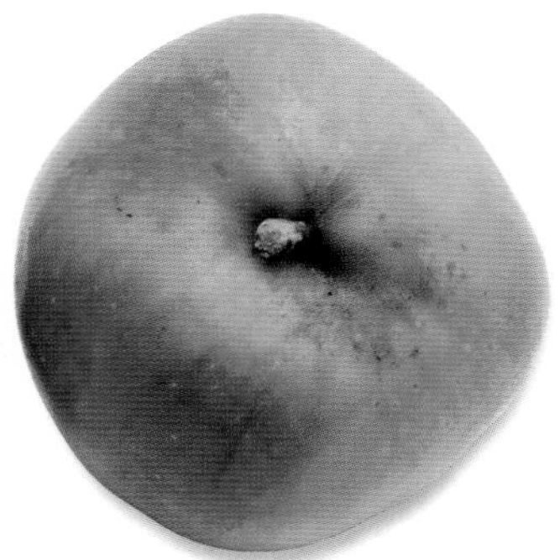

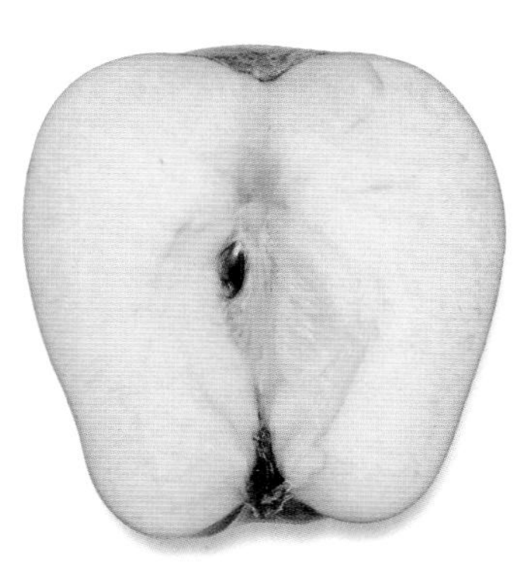

Réaux

The flattened, somewhat irregular fruits, around 6cm (2¼in) across, are bright green with a strong red flush. The creamy white flesh is firm and fine with a sweet, subacid, slightly aromatic flavour.

Type: Dessert.
Origin: France, recorded 1895.
Parentage: Unknown.
Flowering: Late.
Other name: Reau.

Note: This variety is grown in the Meuse, Marne and Argonne departments of France. Fruits can be stored for up to five months.

Red Dougherty

Type: Dessert.
Origin: Twyford, Hawkes Bay, New Zealand, introduced 1930.
Parentage: Unknown.
Flowering: Mid-season.

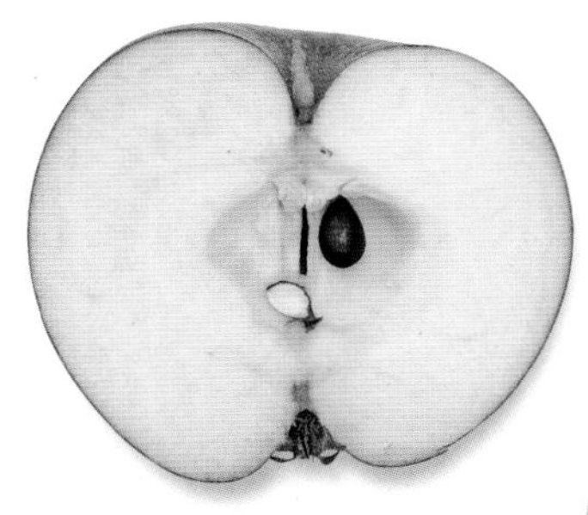

The rounded fruits, around 6cm (2¼in) across, are yellow-green with a pinkish red flush and some striping and flecking. The greenish white flesh is firm, fine and hard with a sweet subacid taste.

Note: This variety has been used in recent breeding programmes to produce apples suitable for dry climates.

Red Granny Smith

Type: Dessert.
Origin: Western Australia, date unknown.
Parentage: ?Granny Smith (female) x ?Jonathan (male).
Flowering: Mid-season.
Other names: Batt's Seedling and Red Gem.

The rounded to slightly conical fruits, around 6cm (2¼in) across, are bright green with a strong red flush and some striping and flecking. The creamy white flesh is firm and crisp with a subacid flavour.

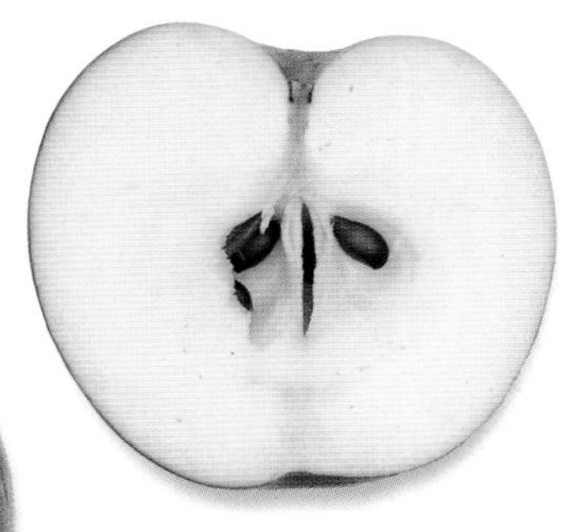

Note: Trees are moderately vigorous. Lady Williams is sometimes grown under this name.

Red Statesman

Type: Dessert.
Origin: New Zealand, discovered 1914.
Parentage: Unknown.
Flowering: Mid-season.
Other name: Warrior.

The rounded fruits, around 6cm (2¼in) across, are bright green with a strong red flush and some striping and spotting. The pale creamy white flesh is firm with a sweet, subacid to acid flavour.

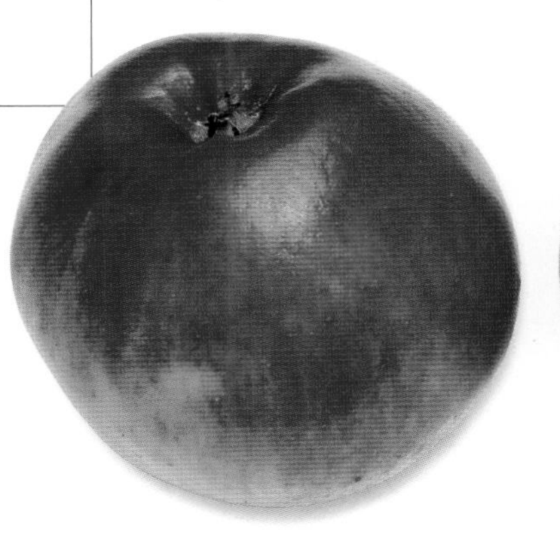

Note: This variety is a more highly coloured clone of the Australian variety Statesman.

Reid's Seedling

The rounded, sometimes slightly irregular or conical fruits, up to around 7cm (2¾in) across, are light green with a carmine red flush and some streaking and russet specks. The creamy white flesh is somewhat coarse with a sweet subacid flavour.

Type: Dessert.
Origin: Richill, Co. Armagh, Ireland, *c.*1880–90.
Parentage: Unknown.
Flowering: Mid-season.

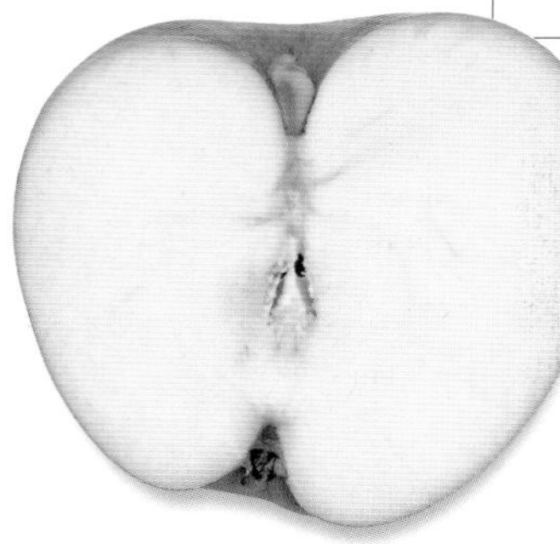

Note: The flesh is sometimes spotted with pink.

Reinette à Longue Queue

Note: This variety is mentioned in *Loudon's Encyclopedia of Gardening* (1822). Fruits have the best flavour eaten soon after picking in summer. They do not store well.

The rounded to conical fruits, around 6cm (2¼in) across or more, are bright yellowish green. The greenish white flesh is firm and rather coarse with a sweetish, slightly aromatic flavour.

Type: Dessert.
Origin: France, recorded 1831.
Parentage: Unknown.
Flowering: Mid-season.
Other names: Reinette à la Longue Queue and Reinette à Longue Quene.

Reinette Clochard

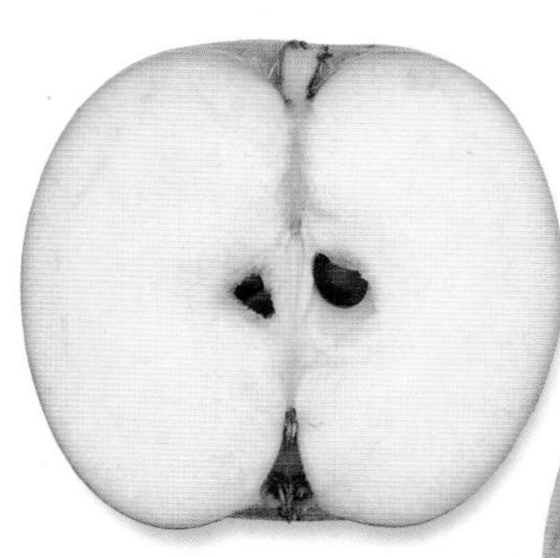

Note: This variety has been grown under many different names, so its lineage is difficult to trace. Trees, of average vigour, crop regularly.

The rounded fruits, around 6cm (2¼in) across, are bright green with some russeting and spotting. The yellowish flesh is fine with a sweet, subacid, perfumed flavour.

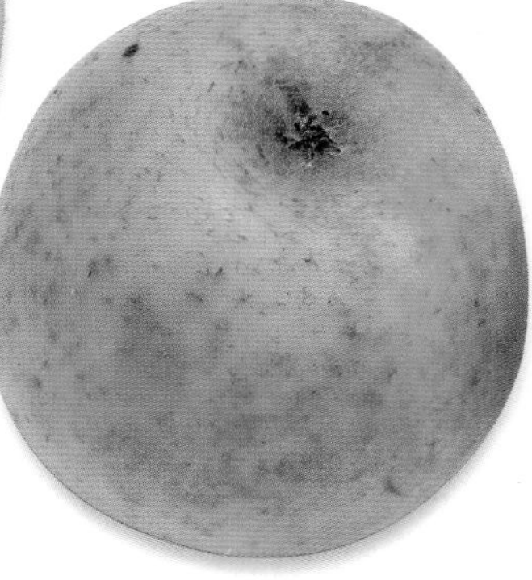

Type: Culinary and dessert.
Origin: France, known mid-1800s.
Parentage: Unknown.
Flowering: Mid-season.
Other names: Clochard, Clochard de Gatine, de Parthenay, Reinette Clocharde, Reinette de Parthenay, Reinette Parthenaise, Reinette von Clochard, Renet Kloshar, Rochelle and Roux brillant.

Reinette d'Anjou

Type: Dessert.
Origin: Belgium or Germany, first mentioned 1817.
Parentage: Unknown.
Flowering: Mid-season.
Other name: Reinette Blanche.

The rounded to slightly conical fruits, around 7cm (2¾in) across, are bright yellow-green with a strong pink to red flush. The creamy white flesh is firm and fine with a subacid, slightly sweet, slightly aromatic flavour.

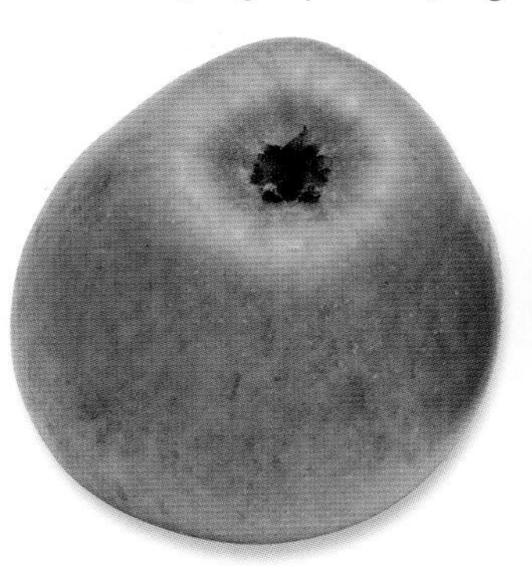

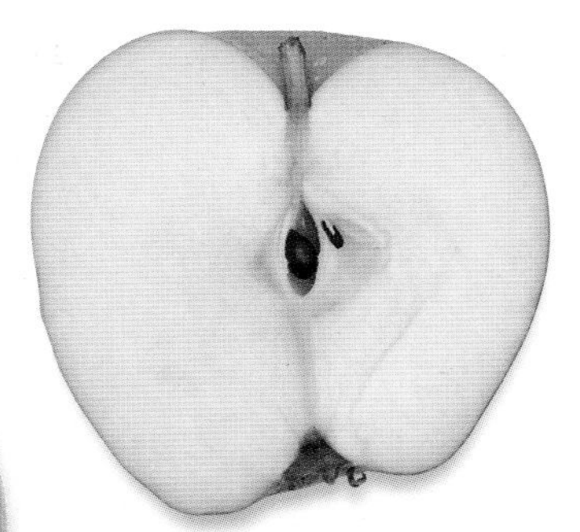

Note: This variety, still grown commercially in France, is similar to Reinette Verte. Trees are hardy, vigorous and bear well. The fruits can be stored for up to five months.

Reinette de Brucbrucks

Type: Dessert.
Origin: France, 1947.
Parentage: Unknown.
Flowering: Mid-season.

The flattened, very irregular fruits, around 7cm (2¾in) across, are green with a strong red flush and some russeting. The creamy white flesh is firm and coarse in texture with a subacid, sweet flavour.

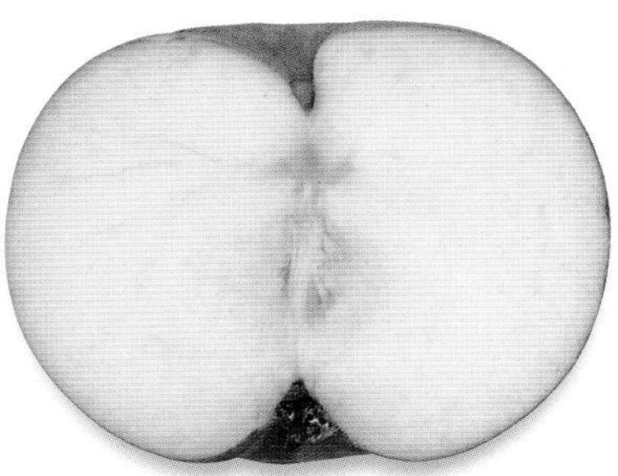

Note: Trees are very vigorous, so are best grown on dwarfing rootstocks.

Reinette de France

Type: Dessert.
Origin: France, 1948.
Parentage: Unknown.
Flowering: Very late.
Other names: Court-pendu de Tournay and Reinette d'Orléans.

The rounded, slightly flattened fruits, up to around 7cm (2¾in) across, are bright yellow-green with a red flush and some russeting. The yellowish white flesh is firm with a subacid, sweet and slightly aromatic flavour.

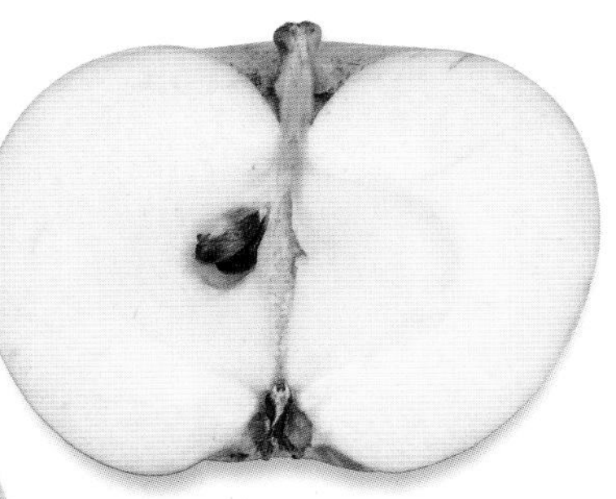

Note: Trees are very hardy and moderately vigorous.

Reinette de Lucas

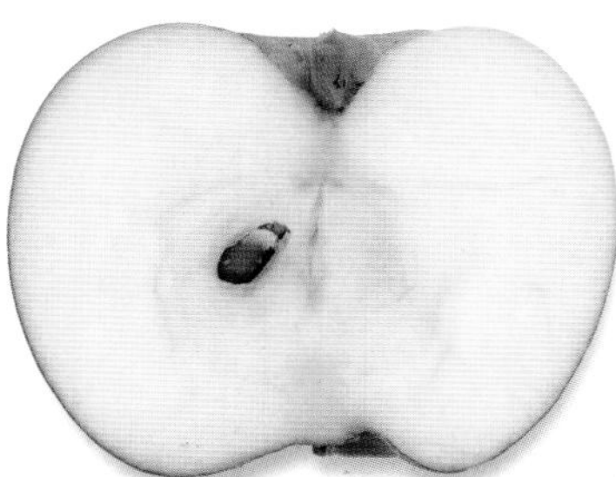

The somewhat flattened, rather irregular fruits, around 7cm (2¾in) across, are bright yellow-green with a strong dark red flush and some russeting. The creamy white flesh is firm, crisp and fine with a subacid and slightly sweet flavour.

Type: Dessert.
Origin: ?Belgium, recorded 1872.
Parentage: Unknown.
Flowering: Mid-season.
Other names: Lucas' Reinette, Lucas's Reinette and Reinette Lucas.

Note: Trees are moderately vigorous.

Reinette de Mâcon

The somewhat flattened, slightly irregular fruits, around 6cm (2¼in) across or more, are dull green with a red flush and dark russeting. The creamy white flesh is firm and fine with a slightly sweet, subacid flavour.

Type: Dessert.
Origin: France, probably Mâcon (Seine-et-Loire), first recorded 1628.
Parentage: Unknown.
Flowering: Early.
Other names: Allman grarenett, Allman laderrenett, Carpentin, Damason Reinette, Double Reinette de Mascon, Laderrenett, Leder Reinette, Leder-Apfel, Lederreinette, Rainette Double de Mazerus, Reinette Damson, Reinette Double de Maserus, Renet Damason, Reneta Damazonska and Styryjskie slynne jablko skorzane.

Note: Early flowering makes this an unsuitable variety for areas where late frosts are likely.

Reinette de Metz

The slightly flattened, somewhat irregular fruits, around 6cm (2¼in) across, are dull yellow-green with a light red flush and some spotting and russeting. The white flesh is fairly fine and tender with a sweet, subacid and slightly aromatic flavour.

Type: Dessert.
Origin: ?France, described 1934.
Parentage: Unknown.
Flowering: Early.

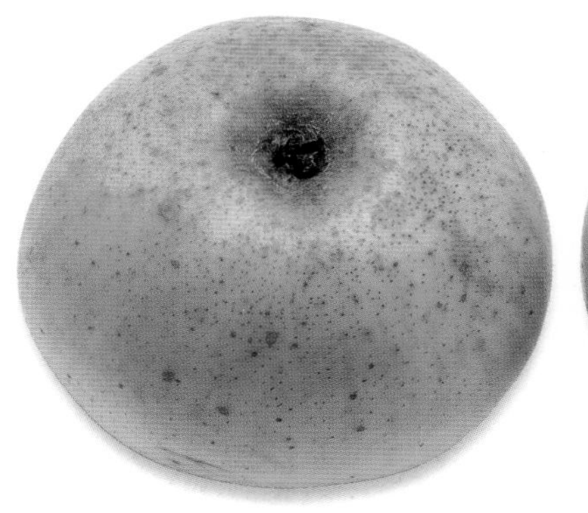

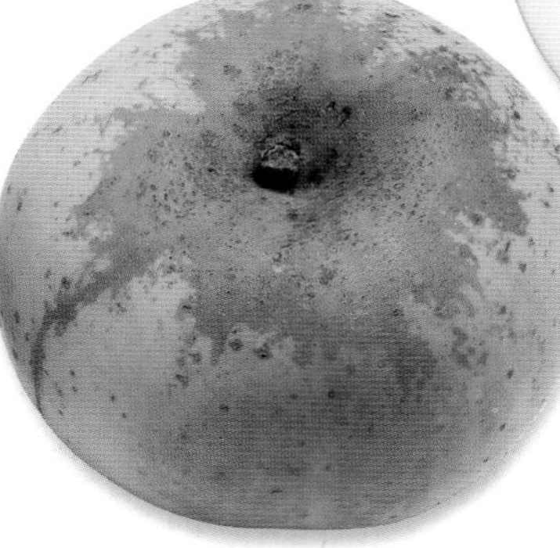

Note: Early flowering makes this an unsuitable variety for areas where late frosts are likely.

Reinette du Canada

Type: Culinary and dessert.
Origin: ?Normandy, France, first mentioned 1771.
Parentage: Unknown.
Flowering: Mid-season.
Other names: Amerikanischer Romanite, Bamporta, Canada Blanc, Canada Pippin, Forbes's Large Portugal, German Green, Hollandische Reinette, Janura, Kaiser-Reinette, Mala Janura, Portugal Russet, Rambour de Paris, Reinette Grandville, Reinette Incomparable, Reinette Virginale, Sternreinette, Wahr Reinette, Weisse Antillische Winter Reinette and White Pippin.

The very irregular, somewhat flattened fruits, around 7cm (2¾in) across, are dull yellow-green with some russeting and spotting. The creamy white flesh is firm, rather dry and coarse-textured with a sweet and moderate flavour.

Note: This variety is a triploid. The trees are vigorous and very productive. The fruits store well.

Reinette Dubuisson

Type: Dessert.
Origin: France, 1950.
Parentage: Unknown.
Flowering: Mid-season.

The somewhat flattened, irregular fruits, up to around 7cm (2¾in) across, are bright yellow-green with a light red flush and some russeting. The greenish white flesh, tinged orange, is firm and coarse with a subacid flavour.

Note: Fruits can be stored for up to six months. Trees are usually very productive and trouble-free.

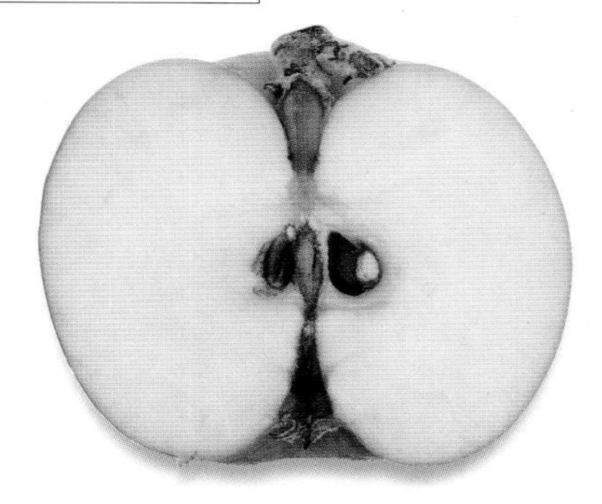

Reinette Marbrée

Type: Dessert.
Origin: Netherlands, described 1760.
Parentage: Unknown.
Flowering: Mid-season.
Other names: Character, Character of Drap d'Or, Charakter Reinette, Cimetière, Concombre des Chartreux, Drap d'Or, Gestrickte Herbst Reinette, Gestrickte Reinette, Heilige Julians Apfel, Julien, Karakter Reinette, Neetjes Apple, Netz Reinette, Pomme de Caractère, Reinette Brodée, Reinette Caractère, Reinette Filée, Reinette Valkenier, Saint Julian and Seigneur d'Orsay.

The rounded, slightly flattened fruits, up to around 7cm (2¾in) across, are dull yellow-green with a marbling of russeting. The whitish flesh is compact, juicy and firm with a very sweet and perfumed flavour.

Note: Trees are moderately vigorous. Fruits can be stored for four months or more.

Reinette Simirenko

The rounded, somewhat irregular fruits, around 6cm (2¼in) across or more, are bright green with some russeting. The greenish white flesh is tender and crisp with a subacid flavour.

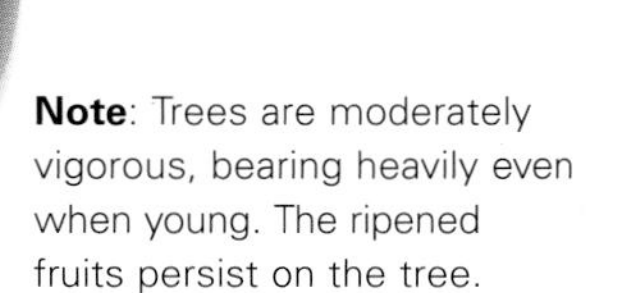

Note: Trees are moderately vigorous, bearing heavily even when young. The ripened fruits persist on the tree.

Type: Dessert.
Origin: Ukraine, described 1895.
Parentage: Unknown.
Flowering: Mid-season.
Other names: Reinette de Simirenko, Reinette Verte de Simirenko, Reinette Verte Incomparable, Renet Filibera, Renet P. F. Simirenko, Renet Simirenko, Simirenko, Simirenkova Reneta, Wood's Greening, Zelenyi renet Simirenko and Zeleomi renet Simirenko.

Reinette Thouin

The rounded to conical fruits, around 6cm (2¼in) across, are bright yellow-green with russeting and spotting. The greenish white flesh is firm, crisp and coarse with a subacid flavour.

Note: Trees are vigorous and productive. Training is possible, but this variety does best when grown as a standard.

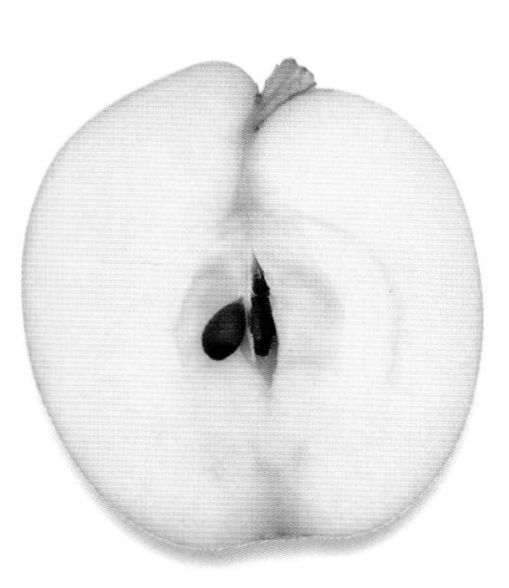

Type: Dessert.
Origin: Beaumont, nr Montmorency, France, 1822.
Parentage: Unknown.
Flowering: Mid-season.
Other names: Bonne Thouin, Renet Tuen, Thouin's Reinette and Thouins Reinette.

Reinette Verte

The rounded, slightly flattened fruits, around 7cm (2¾in) across, are yellowish green with a reddish flush and some greyish russeting and spotting. The yellowish white flesh is tender and juicy with a sweet, vinous, highly aromatic flavour.

Type: Dessert.
Origin: Unknown.
Parentage: Unknown.
Flowering: Late.

Note: The fruits can be stored for up to four months or longer. They are suitable for juicing.

Far right: Fruits develop a very sweet flavour as they ripen.

Ribston Pippin

Type: Dessert.
Origin: Ribston Hall, Yorkshire, UK, from seed brought from Rouen, and planted *c.*1707.
Parentage: Unknown.
Flowering: Early.
Other names: Beautiful Pippin, Formosa, Glory of York, Granatreinette, Jadrnac Ribstonsky, Kaiser Reinette, Lord Raglan, Nonpareille, Pepin de Ribston, Pomme Granite, Poppina Ribston, Reinette de Traves, Reinette Grenade Anglaise, Ribstone, Ribstonov Pepin, Ridge, Rockhill's Russet and Travers.

The slightly irregular fruits, up to around 7cm (2¾in) across, are yellow-green to dull yellow with an orange-red flush and striping and some russeting. The creamy yellow flesh is firm, fine-textured and moderately juicy with a rich aromatic flavour.

Note: This is a probable parent of the Cox's Orange Pippin variety. The flavour is best about one month after picking but fruits can be stored for a further two to three months in optimum conditions. Trees are vigorous and crop well but can be prone to canker in damp soils. Some protection is advisable in cold areas.

Right: Ribston Pippin is an historic variety that is still worth growing.

Richard Delicious

Type: Dessert.
Origin: Unknown.
Parentage: Unknown; presumed to be a sport of Delicious.
Flowering: Mid-season.

The rounded to conical fruits, around 6cm (2¼in) across, are green with an even, bright red flush. The creamy white flesh is sweet and juicy.

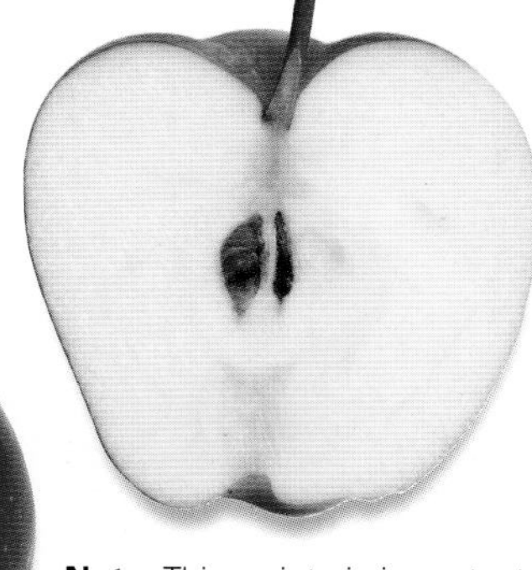

Note: This variety is important commercially in Himachal Pradesh, north India. It is also used in breeding programmes. The fruits can be stored for up to four months.

Roanoke

Type: Dessert.
Origin: Virginia Polytechnic Institute, Blacksburg, USA, introduced 1949.
Parentage: Red Rome (female) x Schoharie (Northern Spy cross) (male).
Flowering: Late.

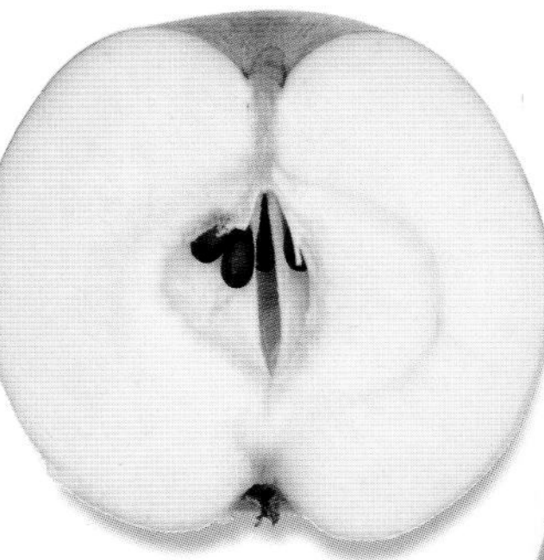

The rounded fruits, around 6cm (2¼in) across, are bright yellow-green with a strong dark red flush. The creamy white flesh is crisp and juicy with a mild flavour.

Note: Trees are moderately vigorous. The name commemorates an important city in Virginia.

Rode Wagenaar

The conical, slightly uneven fruits, around 6cm (2¼in) across, are green with a strong dark red flush. The creamy white flesh is crisp and juicy with a refreshing flavour.

Note: Trees are moderately vigorous.

Type: Dessert.
Origin: IVT, Wageningen, Netherlands, 1963.
Parentage: Unknown.
Flowering: Early.

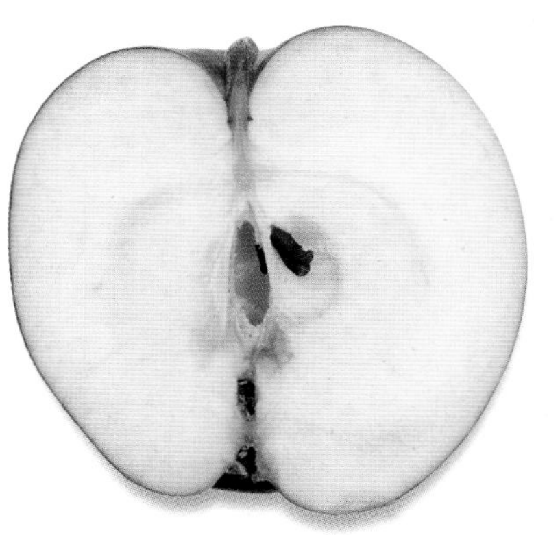

Rosa du Perche

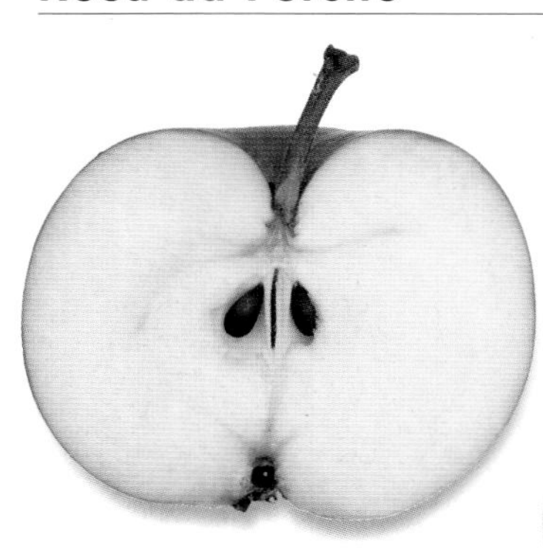

The very uneven, slightly flattened fruits, up to around 7cm (2¾in) across, are green with a strong red flush. The white flesh is fairly fine with a sweet and perfumed flavour.

Type: Dessert.
Origin: France, described 1947.
Parentage: Unknown.
Flowering: Mid-season.
Other name: Pourprée.

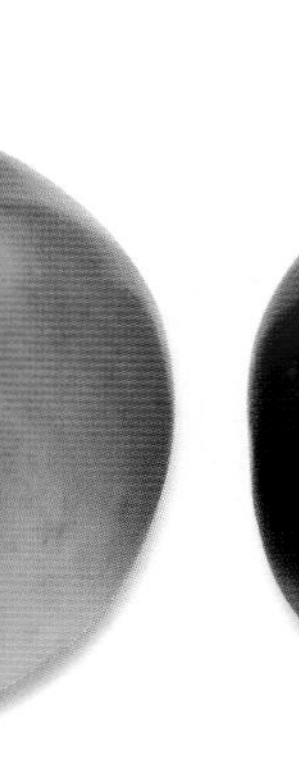

Note: Trees do best in a sheltered site in well-drained soil. The fruits are good for cider making.

Rose de Bénauge

The flattened, slightly uneven fruits, around 7cm (2¾in) across, are bright yellow-green with a strong red flush. The yellowish white flesh is firm and fairly tender with a sweet subacid flavour.

Type: Dessert.
Origin: ?Bordeaux, France, recorded 1872.
Parentage: Unknown.
Flowering: Late.
Other names: Bonne de Mai, Cadillac, de Cadillac, Dieu, Dieudonne, Pomme de Cadillac, Rose (de Knoop), Rose d'Hollande, Rose de Dropt, Rose de Hollande, Rose de la Bénauge, Rose de Mai, Rose du Dropt, Rose Tendre, Rosenapfel aus Benauge and Rozovka iz Benozha.

Note: The fruits are also used in cooking. Late flowering makes this a suitable variety for areas where spring frosts are likely.

Rose de Bouchetière

Type: Dessert.
Origin: France, 1948.
Parentage: Unknown.
Flowering: Early.

Note: This variety is nowadays rare in cultivation.

The flattened fruits, up to around 7cm (2¾in) across or more, are bright green with a netting of russeting. The greenish white flesh is fine but a little tough with a sweet flavour.

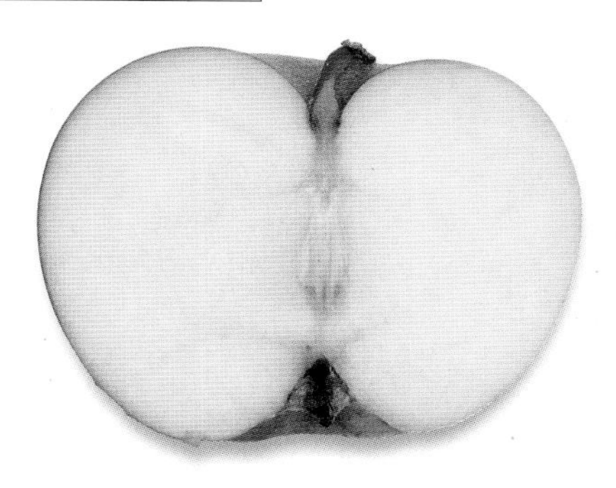

Rose Double

Type: Dessert.
Origin: France, 1948.
Parentage: Unknown.
Flowering: Very late.

The slightly flattened and irregular fruits, around 6cm (2¼in) across, are bright yellowish green with a pinkish red flush and some russeting. The pale cream flesh is firm with a nutty flavour.

Note: Trees are very wind-resistant and are generally free from disease. Young trees may not crop freely.

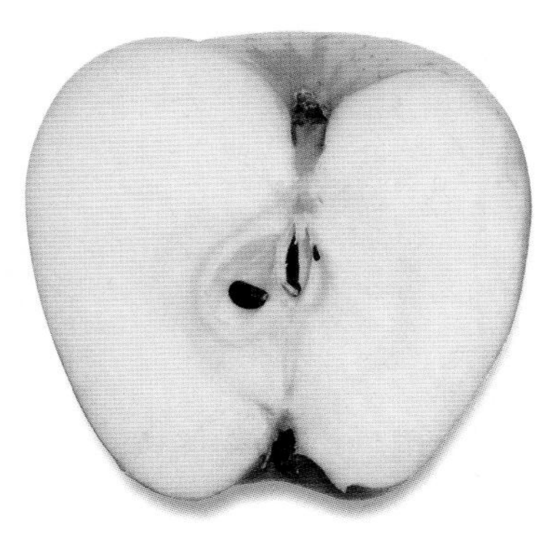

Rose Rouge

Type: Dessert.
Origin: France, 1950.
Parentage: Unknown.
Flowering: Mid-season.

The somewhat flattened, slightly irregular fruits, around 6cm (2¼in) across, are yellow-green with a pinkish red flush and some russeting. The whitish flesh is firm and coarse with a subacid flavour.

Note: This variety is rare in cultivation. Trees are moderately vigorous.

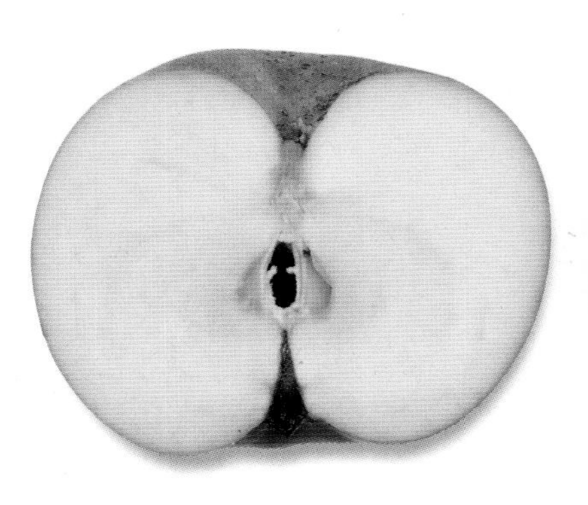

Rosemary Russet

The somewhat conical, irregular fruits, up to around 7cm (2¾in) across or more, are yellow-green with a red flush and some pale brown russeting. The creamy white flesh is firm, fine-textured and juicy with a rather acid and good flavour.

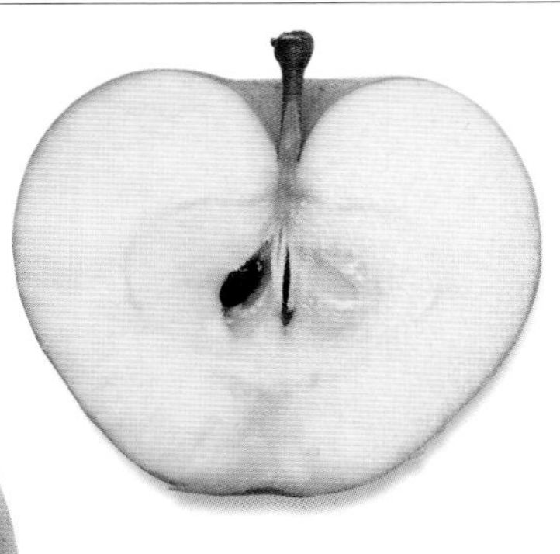

Type: Dessert.
Origin: England, UK, first described 1831.
Parentage: Unknown.
Flowering: Mid-season.
Other names: Benskin's Russet, Buzzan, Rosemary and Rosemary Apple.

Note: The fruits can be stored for two to three months. Trees are moderately vigorous.

Ross Nonpareil

The rounded fruits, around 6cm (2¼in) across, are green with a light flush and a netting of russeting. The creamy white to greenish flesh is firm and rather dry with a rich, aromatic flavour.

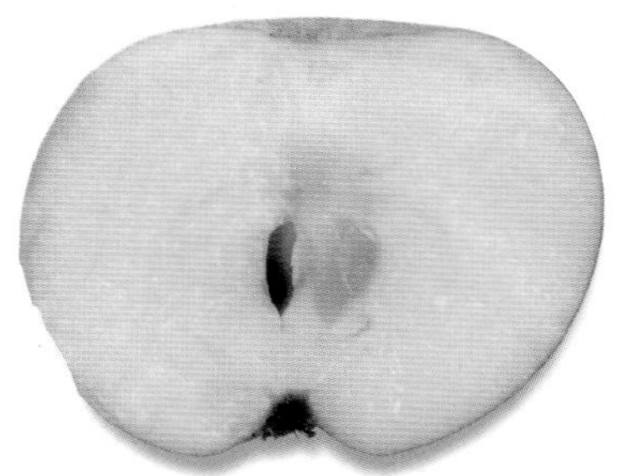

Left: Fruits of this variety are generally russeted.

Type: Dessert.
Origin: Ireland, UK, recorded 1802, introduced to England 1819.
Parentage: Unknown.
Flowering: Early.
Other names: French Pippin, Lawson Pearmain, Non-Pareille de Ross, Nonpareil Ross and Nonpareille de Ross.

Note: The trees are very hardy and are tolerant of most soils. They bear freely. The fruits can be stored for two to three months.

Rossie Pippin

The flattened, irregular fruits, up to around 7cm (2¾in) across or more, are an even bright green. The greenish white flesh is fine with an acid flavour.

Type: Culinary.
Origin: Maidstone, Kent, UK, catalogued 1890.
Parentage: Unknown.
Flowering: Mid-season.

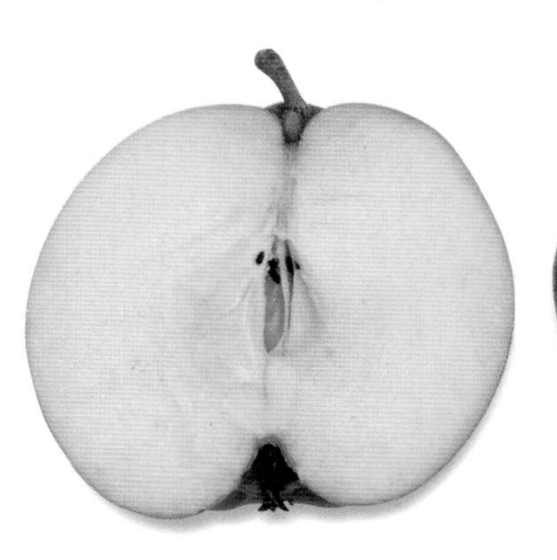

Note: This variety is rare in cultivation.

Right: The fruits retain an even green colouring.

Roter Eiserapfel

Type: Dessert.
Origin: Germany, early 1700s.
Parentage: Unknown.
Flowering: Mid-season.
Other names: Arsapple, Bamberger, Cristapfel, Durable Trois Ans, Eiser Rouge, Fragone, Hunt's Royal Red, Jarnapple, Kloster Apfel, Mohrenkopf, Nagelsapfel, Red Eisen, Roter Bach, Roter Jahrapfel, Roter Krieger, Roter Paradiesapfel, Rother Backapfel, Rother Eiser, Rott Jarnapple, Rouge Ravée, Schorsteinfeger, Tartos Piros Alma, Treckhletnee, Zelezne jablko and Zelezniac.

The rounded to conical fruits, around 7cm (2¾in) across, are yellow-green with a strong dark red flush. The creamy white flesh is very hard and fine with a sweet, subacid flavour.

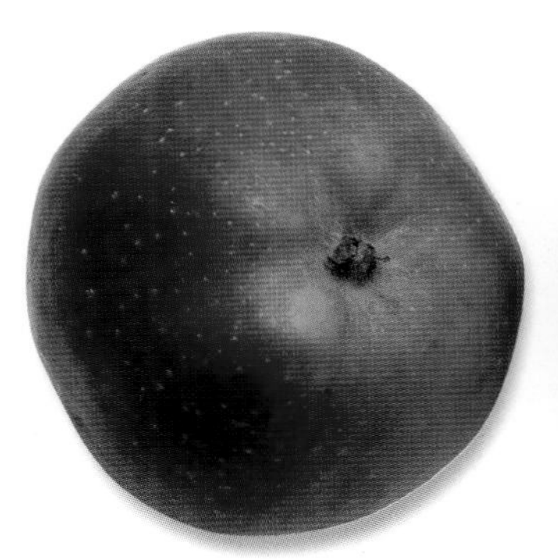

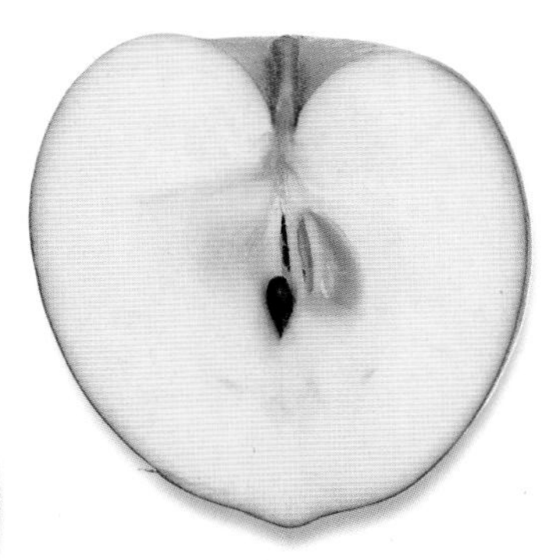

Note: These vigorous trees crop heavily but are prone to biennial bearing. The flowers are resistant to frost. The fruits are also suitable for cooking.

Roter Münsterländer Borsdorfer

Type: Dessert.
Origin: Germany, 1951.
Parentage: Unknown.
Flowering: Mid-season.

The rounded, slightly irregular fruits, around 6cm (2¼in) across or more, are bright green with a strong dark red flush and some russeting. The greenish white flesh is firm and somewhat coarse with a subacid and slightly sweet flavour.

Note: Fruits can be stored for two to three months. Trees are generally disease resistant.

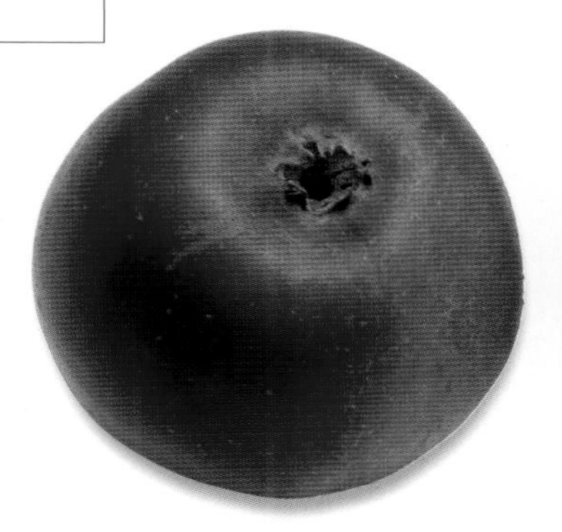

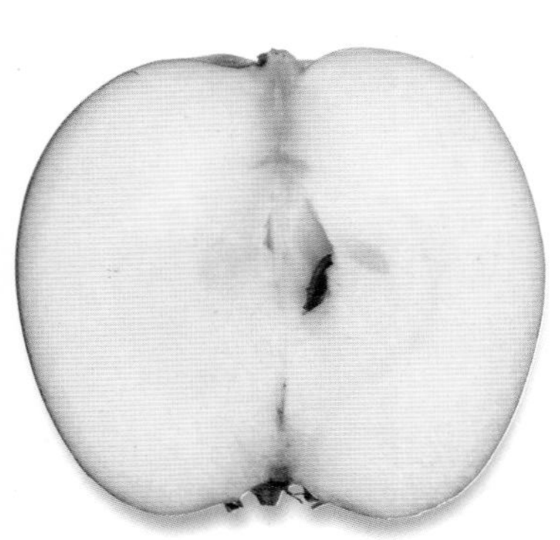

Roter Sauergrauech

Type: Dessert.
Origin: Switzerland, 1947.
Parentage: Unknown.
Flowering: Early.
Other name: Sauergrauech Rouge.

The rounded to slightly conical fruits, around 6cm (2¼in) across, are bright yellow-green with a strong red flush and some streaking and striping. The greenish white flesh is fine and soft with a slightly sweet, vinous, aromatic flavour.

Note: This variety is a more highly coloured sport of Sauergrauech.

Roter Stettiner

The irregular fruits, sometimes slightly flattened, around 6cm (2¼in) across or more, are bright yellow-green with a deep pink to red flush. The greenish white flesh is firm and fine with a subacid, sweet flavour.

Note: Trees are moderately vigorous. This variety does well in cool areas. The flowers are frost resistant. The fruits can be stored for several months.

Type: Dessert.
Origin: Germany, first described 1598.
Parentage: Unknown.
Flowering: Early.
Other names: Adam, Annaberger, Belle Hervey, Botzen, Butter-Apfel, Calviller, d'Hiver, Eisenapfel, Guly-Muly, Herrenapfel, Kack, Kohl, Mahler, Matapfel, Paradies Apfel, Pomme de Fer Vineuse, Rouge de Stettin, Rubiner, Schuller, Schwer, Shchetinka, Stetting Rouge, Tragamoner, Vejlimek chocholaty, Vineuse Rouge, Winter Sussapfel, Wintersuss, Wittlaboth and Zwiebelapfel.

Roxbury Russet

The very irregular fruits, up to around 7cm (2¾in) across or more, are bright yellowish green and greenish bronze with some spotting. The creamy white flesh is somewhat coarse and fairly tender with a subacid flavour.

Note: This is possibly the oldest American variety. The fruits store well.

Type: Dessert.
Origin: Roxbury, MA, USA, early 1600s.
Parentage: Unknown.
Flowering: Mid-season.
Other names: Belper Russet, Boston Russet, English Russet, Hewe's Russet, Howe's Russet, Jusset, Marietta Russet, Pitman's Russet, Pomme Russet, Putman Russet, Putnam Russet, Putnam Russet of Ohio, Renet Bostonskii, Rox, Rox Russet, Roxburg Russet, Roxbury Russeting, Shippen's Russet, Sylvan Russet and Warner's Russet.

Royal Gala

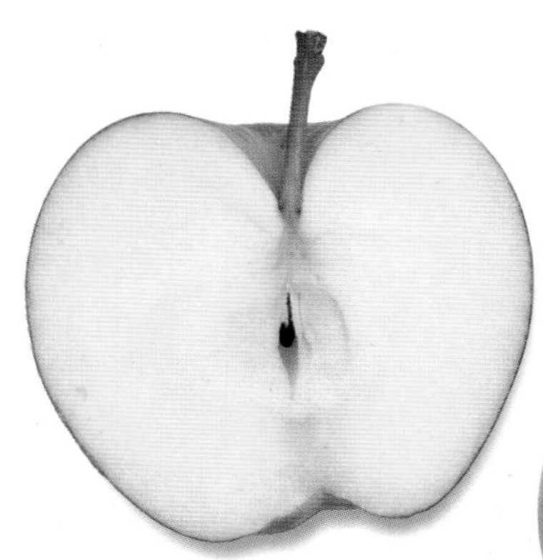

Note: This is a very important commercial variety. Trees are moderately vigorous.

The rounded fruits, around 7cm (2¾in) across, are green with a pinkish red flush. The creamy white flesh is firm, crisp, fine-textured and juicy with a sweet and good aromatic flavour.

Type: Dessert.
Origin: A cultigen made from a sport of Gala, 1970s.
Parentage: Kidd's Orange Red (female) x Golden Delicious (male).
Flowering: Early.

Runse

Type: Dessert.
Origin: Italy, 1958.
Parentage: Unknown.
Flowering: Mid-season.

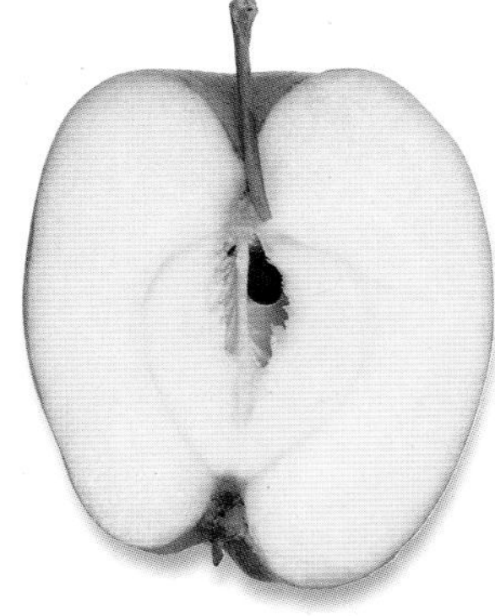

Note: This variety is rare in cultivation.

Left: Fruits develop red striping on the side facing the sun.

The rounded to slightly conical fruits, around 6cm (2¼in) across, are bright yellow-green with a strong red flush and some striping. The yellowish white flesh is soft and coarse with a subacid flavour.

Sacramentsappel

Type: Dessert.
Origin: Plant Breeding Institute, Wageningen, Netherlands, 1955.
Parentage: Unknown.
Flowering: Very late.

The very irregular fruits, around 7cm (2¾in) across, are yellow with a strong red flush. The creamy white flesh is fine and tender with a subacid and slightly rich flavour.

Note: Late flowering makes this variety suitable for areas where late spring frosts are likely.

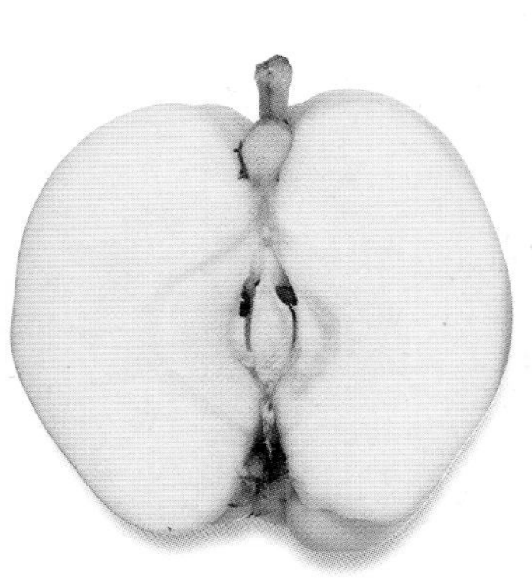

Saint Edmund's Pippin

Type: Dessert.
Origin: Bury St Edmunds, Suffolk, UK, recorded 1875.
Parentage: Unknown.
Flowering: Early.
Other names: Early Golden Russet, St Edmonds, St Edmund's Pippin, St Edmund's Russet and St Edmunds.

The rounded fruits, up to around 7cm (2¾in) across or more, are bright green with extensive greenish brown russeting that makes the skins very rough. The creamy yellow flesh is moderately firm, juicy and slightly acid with a good flavour.

Note: This variety tolerates a range of growing conditions, is neat growing and shows good disease resistance. The fruits, which tend to shrivel in storage, are particularly good for juicing and cider making.

Right: Fruits develop a rough, sandpaper-like skin that is typical of russet apples.

Salome

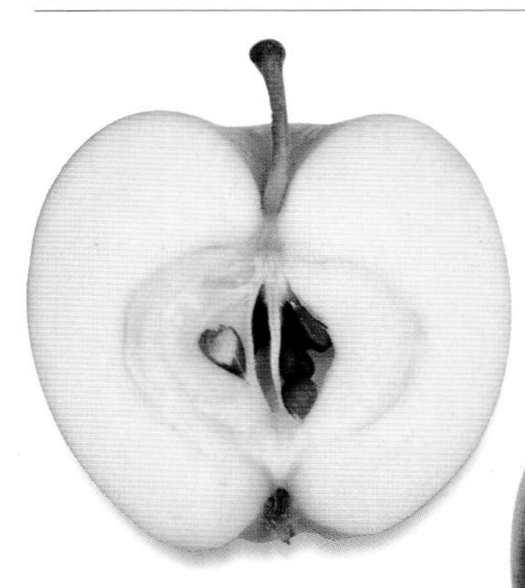

The rounded to conical, slightly uneven fruits, around 6cm (2¼in) across, are yellow-green with a light pinkish red mottled flush and flecking. The greenish white flesh is firm and fine with a subacid flavour.

Type: Dessert.
Origin: Ottawa, IL, USA, introduced 1884.
Parentage: Unknown.
Flowering: Early.

Note: The fruits can be ribbed. Trees are moderately vigorous. They are hardy and crop abundantly.

Sam Young

The flattened fruits, around 5cm (2in) across, are bright green with a bright red flush and are almost covered with grey russeting. The greenish white flesh is firm and rather tough with a subacid flavour.

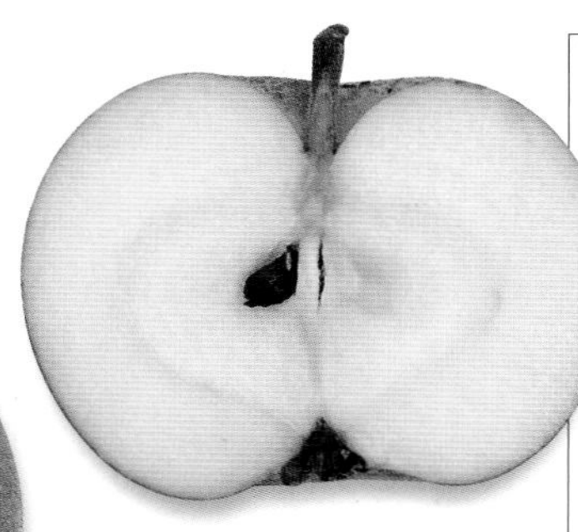

Type: Dessert.
Origin: Ireland, brought to notice in England, UK, 1818.
Parentage: Unknown.
Flowering: Early.
Other names: Irish Russet, Irlandischer Rothling and Irlandischer Rotling.

Note: Trees are moderately vigorous. The fruits can be stored for two to three months.

Sandlin Duchess

The flattened fruits, often more than 7cm (2¾in) across, are bright green with some flushing and russeting. The creamy green flesh is fine and tender with a subacid and slightly sweet flavour.

Type: Dessert.
Origin: Sandlin, Malvern, Worcestershire, UK, 1880.
Parentage: Unknown.
Flowering: Mid-season.

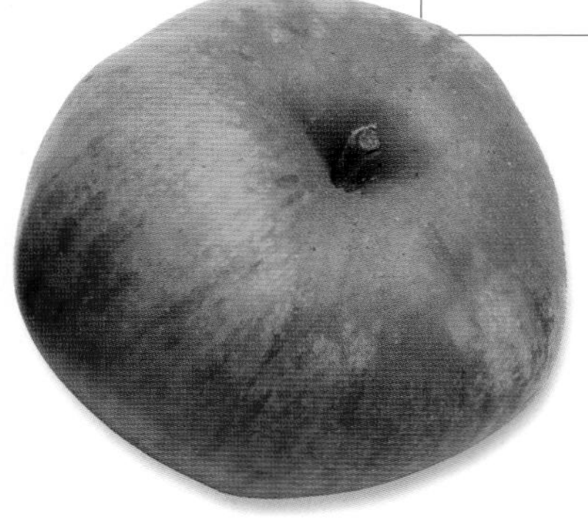

Note: Fruits store well for around four months. Newton Wonder has been claimed to be in its parentage.

Sandow

Type: Dessert.
Origin: Central Experiment Farm, Ottawa, Canada, selected 1912, introduced 1935.
Parentage: Northern Spy (female) x Unknown.
Flowering: Mid-season.

The rounded, slightly flattened fruits, more than 6cm (2¼in) across, are bright yellow-green with a strong dark red flush and striping. The creamy white flesh is coarse and tender with a sweet to fairly acid, and sometimes raspberry-like, flavour.

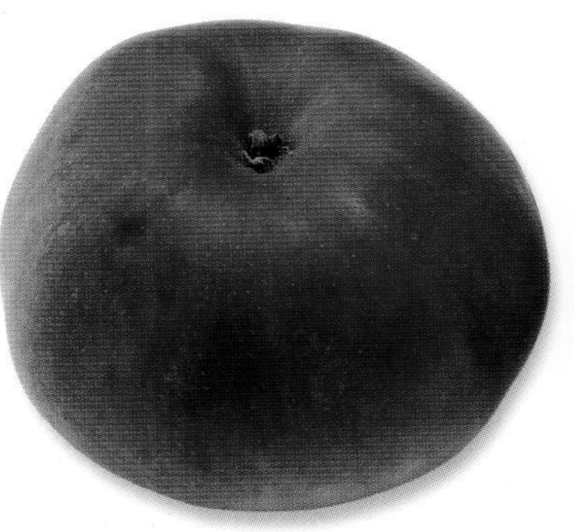

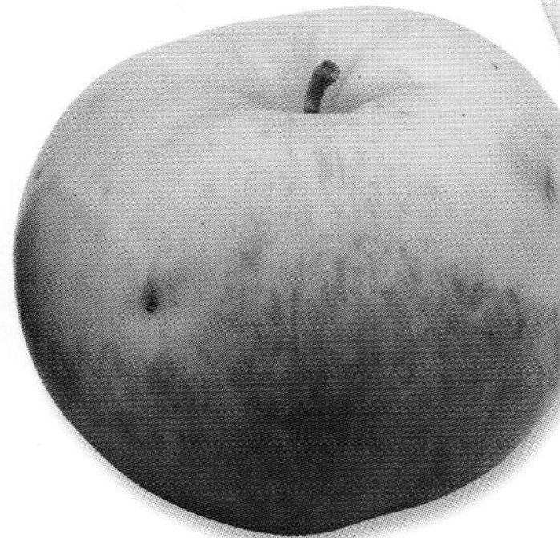

Note: This variety is similar to its parent Northern Spy but is hardier and less prone to scab. The fruits have a fuller flavour.

Sanspareil

Type: Dessert.
Origin: Known in England, UK, since the late 1800s.
Parentage: Unknown.
Flowering: Early.

The rounded to slightly flattened fruits, around 6cm (2¼in) across, are bright green with a pinkish red flush and a small amount of russeting. The yellow flesh is crisp with a sweet, aromatic flavour.

Note: The fruits can be stored for several months. Trees are moderately vigorous.

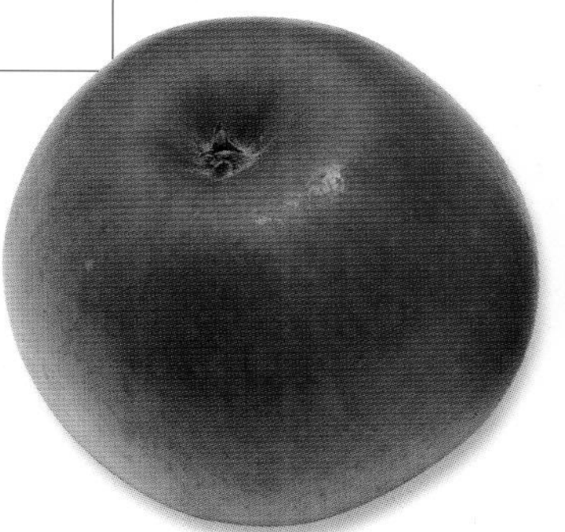

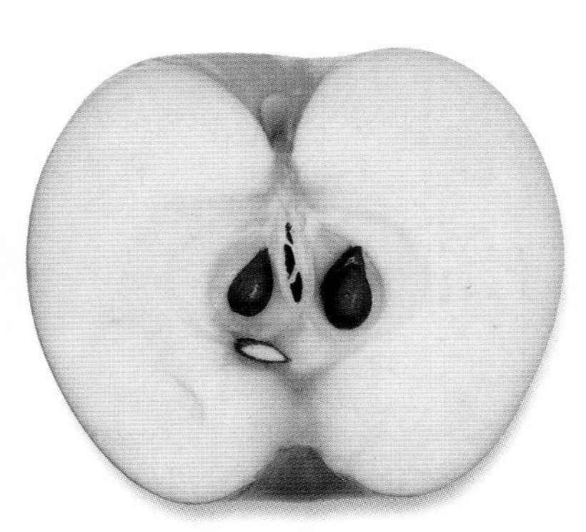

Schöner aus Herrnhut

Type: Dessert.
Origin: Herrnhut, Germany.
Parentage: Unknown.
Flowering: Early.
Other names: Herrnhutsk and Piekna z Herrnhut.

Note: This variety is moderately susceptible to scab and mildew. Biennial bearing can be a problem. Fruits are good for juicing.

The rounded to irregular fruits, up to around 7cm (2¾in) across or more, are bright yellow-green with a strong red flush and some striping and spotting. The yellowish white flesh is juicy with a sweet but tart flavour.

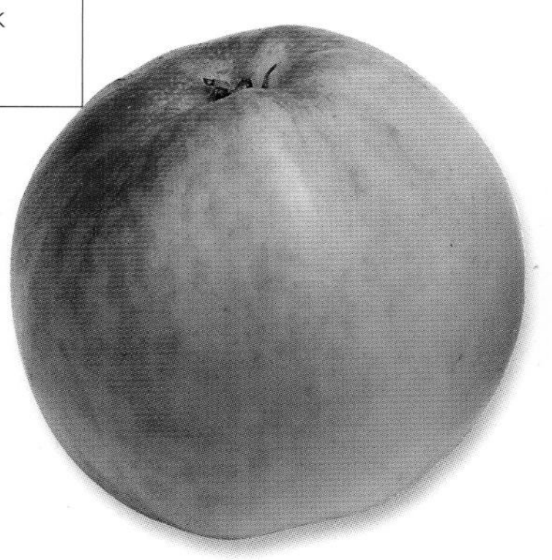

Above: The fruits of this variety can vary in shape, and develop their red colouring on the side facing the sun.

Schweizer Orange

The flattened, uneven fruits, around 7cm (2¾in) across or more, are bright yellow-green with a strong red flush. The creamy white flesh is firm with a subacid flavour.

Type: Dessert.
Origin: Swiss Federal Agricultural Research Station, Wädenswil, Switzerland, 1935, released 1955.
Parentage: Ontario (female) x Cox's Orange Pippin (male).
Flowering: Mid-season.
Other names: Schweizer Orangenapfel and Suisse Orange.

Note: This variety has a tendency to produce quantities of small fruits.

September Beauty

The rounded, slightly irregular fruits, around 7cm (2¾in) across, are bright green with a bright red flush and some russeting. The creamy white flesh is coarse and loose with a fairly sweet flavour.

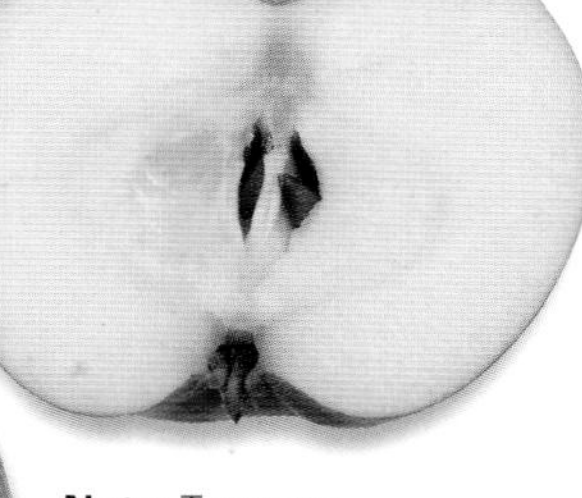

Type: Dessert.
Origin: Bedford, UK, 1885.
Parentage: Unknown.
Flowering: Mid-season.

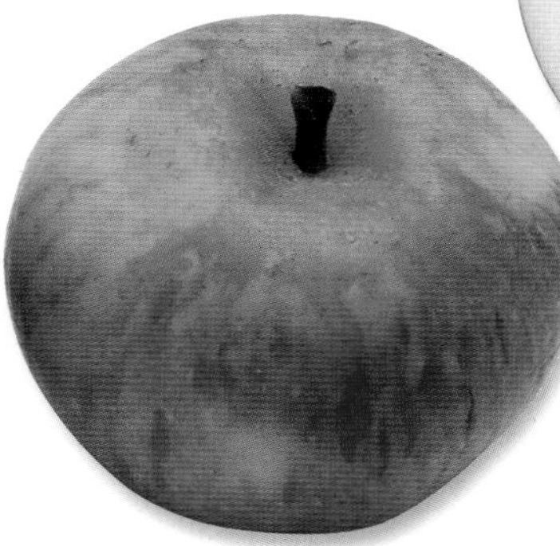

Note: Trees are moderately vigorous.

Right: This variety was raised by Laxtons, a noted Victorian grower.

Sharon

The rounded fruits, around 6cm (2¼in) across, are bright green with a strong red flush and some flecking and striping. The whitish green flesh is juicy, fine and tender with a sweet flavour.

Note: The flavour is similar to McIntosh but sweeter and firmer.

Type: Dessert.
Origin: Iowa State Agricultural Experiment Station, Ames, USA, 1906, introduced 1922.
Parentage: McIntosh (female) x Longfield (male).
Flowering: Early.

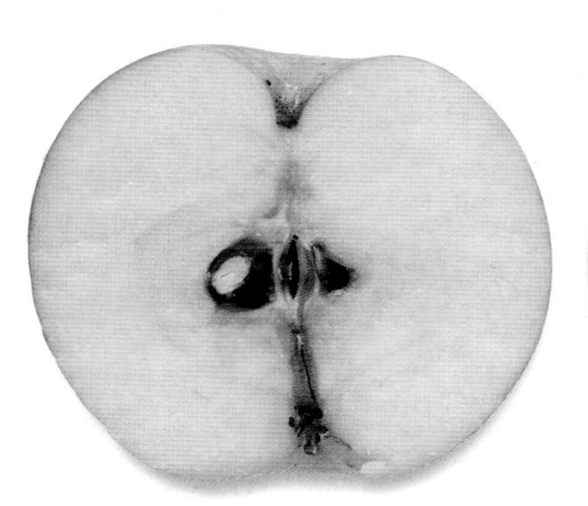

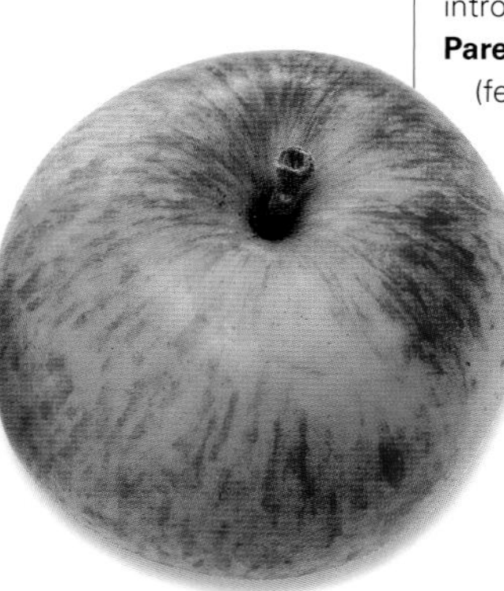

Shin Indo

Type: Dessert.
Origin: Aomori Apple Experiment Station, Japan, raised 1930, named 1948.
Parentage: Indo (female) x Golden Delicious (male).
Flowering: Mid-season.

The conical fruits, around 7cm (2¾in) across or more, are bright green with a pinkish red flush. The cream flesh, tinged green, is firm and dry with a sweet flavour.

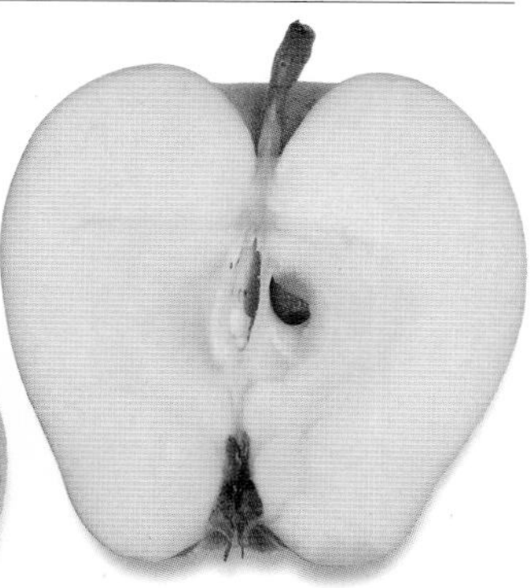

Note: Trees are very vigorous, so should be grown on dwarfing rootstocks.

Siddington Russet

Type: Culinary and dessert.
Origin: Siddington, Gloucestershire, UK, 1923.
Parentage: Unknown.
Flowering: Mid-season.

Note: This variety, currently endangered, is a russeted sport of Galloway Pippin.

The flattened fruits, around 6cm (2¼in) across, are bright green with a netting of russeting. The creamy white flesh is firm, crisp and juicy with a subacid flavour.

Signe Tillisch

Type: Culinary and dessert.
Origin: ?Denmark, *c.*1866, first described 1889.
Parentage: ?Calville Blanc D'Hiver (female) x Unknown.
Flowering: Mid-season.
Other names: Signe Tillish and Sini Tillis.

The rounded fruits are bright green with a bright red flush and some striping. The creamy white flesh is loose and soft with a sweet, subacid, aromatic flavour.

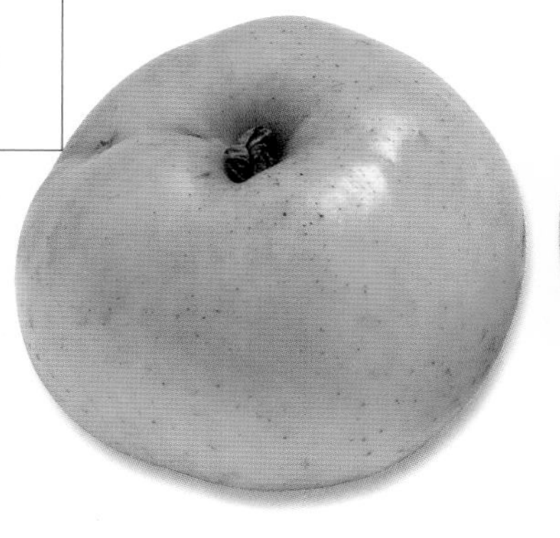

Note: This variety was raised by Councillor Tillisch and named after his daughter. Trees are moderately vigorous.

Sikulai-alma

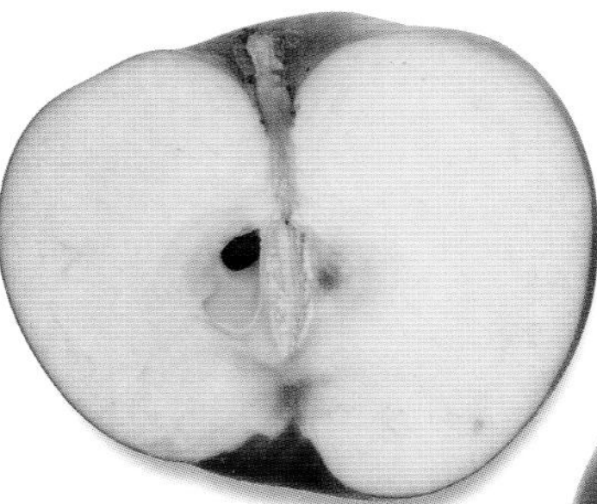

Note: This variety is resistant to fireblight. It is self-sterile, so a suitable pollinator is required for good fruiting.

The irregular, somewhat flattened fruits, around 7cm (2¾in) across, are bright green with a strong dark red flush. The greenish white to cream flesh is firm and fine with a subacid flavour.

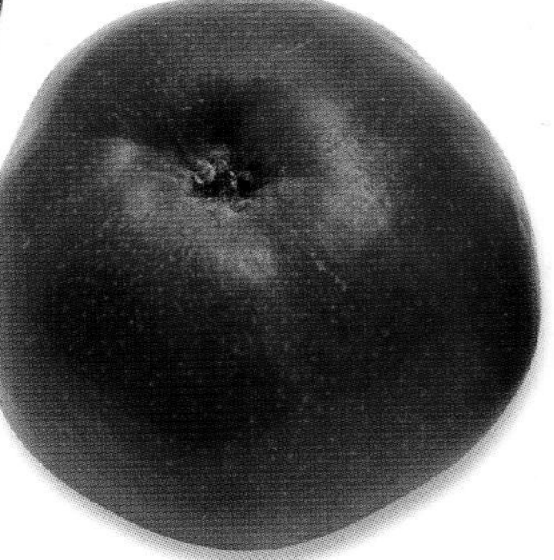

Type: Dessert.
Origin: Hungary, first recorded 1875.
Parentage: Unknown.
Flowering: Mid-season.
Other names: de Sikula, Seklerapfel, Siculane, Sikula, Sikulaer Apfel, Sikulaerapfel, Sikulai Alma, Sikulaske, Szekely-alma and Szekelyalma.

Simonffy Piros

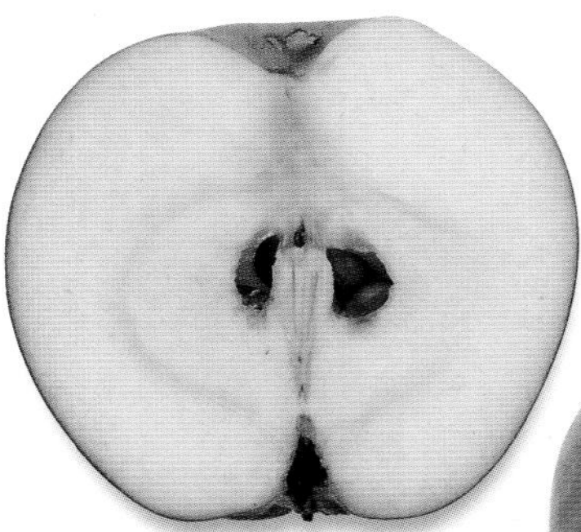

Note: This variety is resistant to fireblight.

The rounded fruits, around 7cm (2¾in) across or more, are bright yellow-green with a strong dark red flush and some russeting. The greenish white flesh is firm and fine with a very sweet, subacid, perfumed, strawberry-like taste.

Type: Dessert.
Origin: Hungary, recorded 1876.
Parentage: Unknown.
Flowering: Early.
Other name: Simonffy Roth.

Sir John Thornycroft

The flattened fruits, around 7cm (2¾in) across, are bright yellow-green with a red flesh, some red flecking and some russeting. The creamy white flesh is hard, tough and coarse with a slightly sweet flavour.

Note: This variety is now rare in cultivation but is often included in heritage plantings.

Type: Dessert.
Origin: Bembridge, Isle of Wight, UK, introduced 1913.
Parentage: Unknown.
Flowering: Mid-season.
Other name: Sir John Thorneycroft.

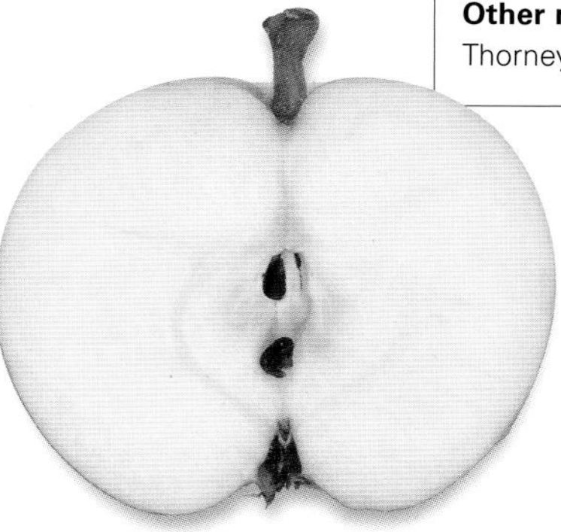

Sisson's Worksop Newton

Type: Dessert.
Origin: Worksop, Nottinghamshire, UK, 1910.
Parentage: Newtown Pippin (female) x Unknown.
Flowering: Mid-season.

The irregular, somewhat flattened-to-square fruits, up to around 7cm (2¾in) across, are pale greenish-yellow with a light orange-brown flush and some light spotting. The greenish white flesh is coarse, with a subacid to acid flavour.

Note: This variety needs a sunny location in order for the flavour to develop its full pineapple-like aroma.

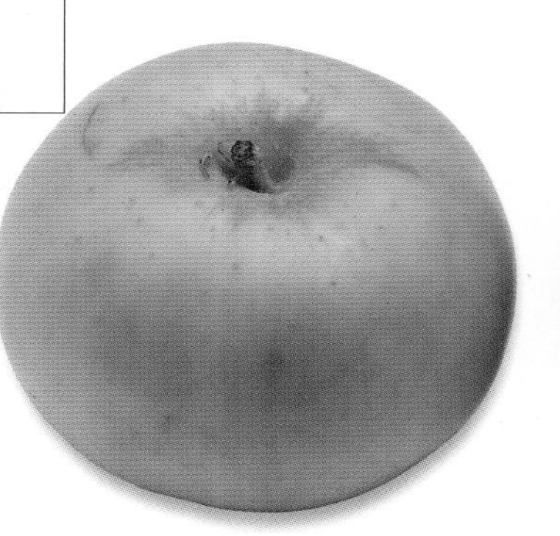

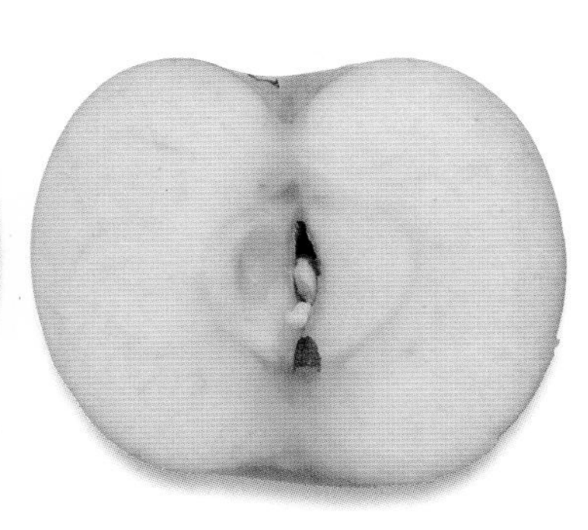

Slavyanka

Type: Dessert.
Origin: Russia, 1899.
Parentage: Antonovka (female) x Ananas Reinette (male).
Flowering: Mid-season.
Other names: Slavianka and Slavjanka.

The rounded, irregular fruits, around 6cm (2¼in) across, are an even bright yellow-green. The white flesh is fine with a sweet, subacid, aromatic flavour.

Note: This variety is a partial tip-bearer.

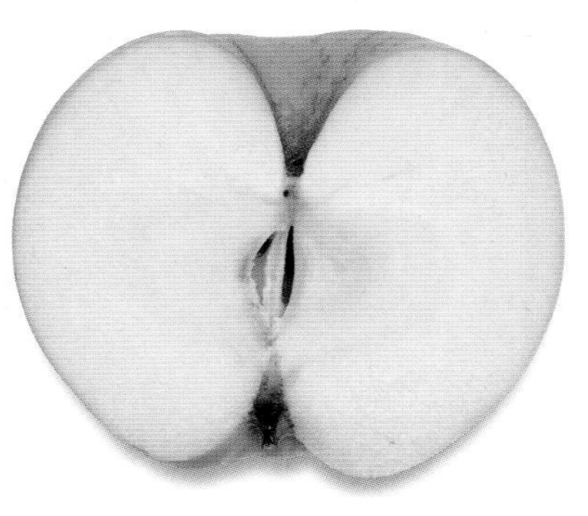

South Park

Type: Dessert.
Origin: South Park, Penshurst, Kent, UK, 1940.
Parentage: Cox's Orange Pippin (female) x Winter Queening (male).
Flowering: Mid-season.

The rounded fruits, around 6cm (2¼in) across or more, are bright yellow-green with a strong red flush and some russeting. The cream flesh, tinged green, is crisp with an acid flavour.

Note: Trees are moderately vigorous.

Spencer Seedless

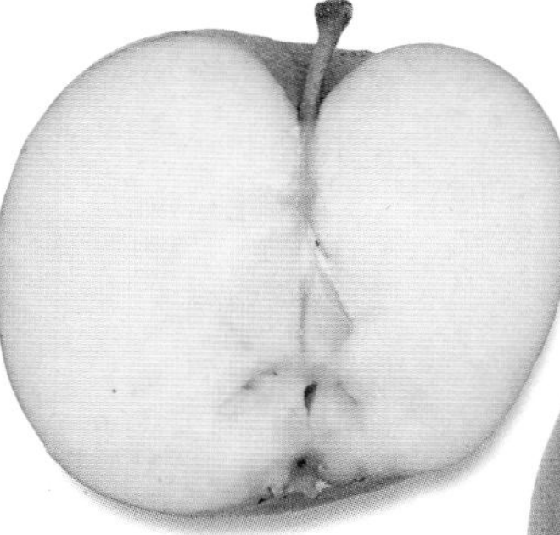

The rounded or slightly flattened fruits, around 6cm (2¼in) across, are dull yellowish green with a red flush and russeting. The creamy white flesh has poor flavour.

Type: Dessert.
Origin: Long Ashton Research Station, Bristol, UK, 1970.
Parentage: Unknown.
Flowering: Late.

Note: The fruits of this variety rarely set any seed.

Stark's Late Delicious

The conical, irregular fruits, around 7cm (2¾in) across, are bright yellow-green with a pinkish red flush and some striping. The cream flesh, tinged green, is very sweet.

Type: Dessert.
Origin: Scotland, UK, 1967.
Parentage: Unknown.
Flowering: Mid-season.

Note: Trees are moderately vigorous.

Right: Fruits of this variety taper sharply towards the blossom end.

Starkrimson Delicious

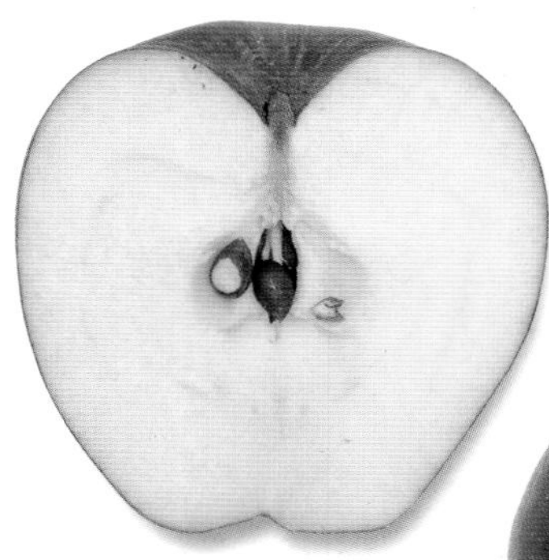

The conical fruits, around 6cm (2¼in) across, are bright green with a strong dark red flush. The creamy white flesh is firm, very sweet and juicy with a highly aromatic flavour.

Type: Dessert.
Origin: Hood River, OR, USA, *c.*1953, introduced 1956.
Parentage: Unknown.
Flowering: Mid-season.
Other names: Bisbee Red Delicious and Starkrimson.

Note: This variety, which can be weak-growing, is a more highly coloured and taller-fruited clone of Starking. Trees bear well and are resistant to fungal diseases.

Stearns

Type: Dessert.
Origin: North Syracuse, NY, USA, recorded 1900.
Parentage: Esopus Spitzenburg (female) x Unknown.
Flowering: Mid-season.

The slightly irregular fruits, up to 8cm (3in) across, are bright green with a pinkish red flush and some striping. The creamy white flesh is crisp yet melts in the mouth with a sweet, subacid flavour.

Note: Trees are very vigorous, so are suitable for growing on dwarfing rootstocks.

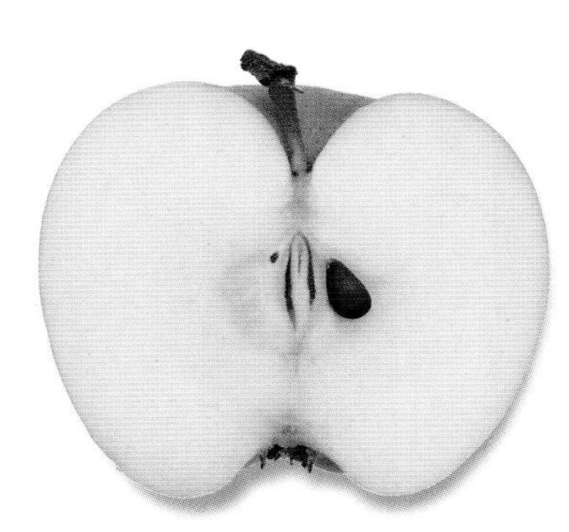

Steyne Seedling

Type: Dessert.
Origin: Steyne, Isle of Wight, UK, *c.*1893.
Parentage: Unknown.
Flowering: Early.
Other name: Steyne's Seedling.

The flattened fruits, around 6cm (2¼in) across or more, are bright yellow-green with a pinkish red flush and some streaking. The creamy white flesh is soft, tender and juicy with a subacid flavour.

Note: This is one of several notable varieties raised in the gardens of Sir John Thornycroft on the Isle of Wight. The flavour is similar to Cox's Orange Pippin.

Stina Lohmann

Type: Dessert.
Origin: Germany, 1951.
Parentage: Unknown.
Flowering: Mid-season.

The rounded but irregular fruits, around 6cm (2¼in) across, are bright green with a red flush and some russeting and streaking. The yellowish white flesh is firm and fine with a slightly sweet, subacid, somewhat rich flavour.

Note: Trees are vigorous and tolerant of a range of soil types. Fruits can be stored for up to six months or even longer.

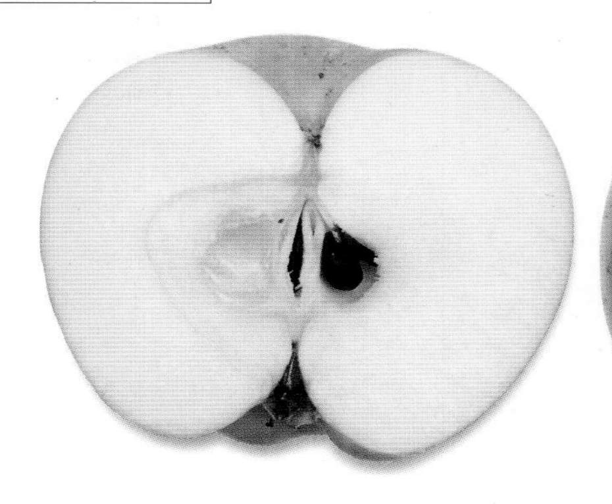

Stoke Edith Pippin

The rounded, slightly irregular fruits, around 6cm (2¼in) across or more, are bright yellow-green, tinged orange, with some grey russeting and spotting. The yellow flesh is firm and crisp with a subacid, sweet and perfumed flavour.

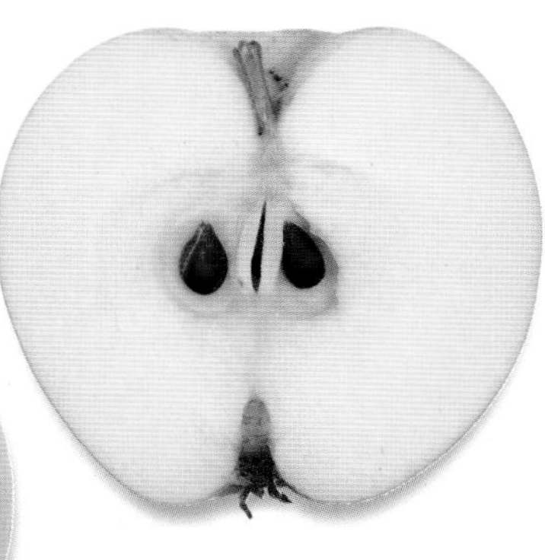

Type: Dessert.
Origin: ?Stoke Edith, Herefordshire, UK, recorded 1872.
Parentage: Unknown.
Flowering: Mid-season.
Other name: Stock-Edith Pippin.

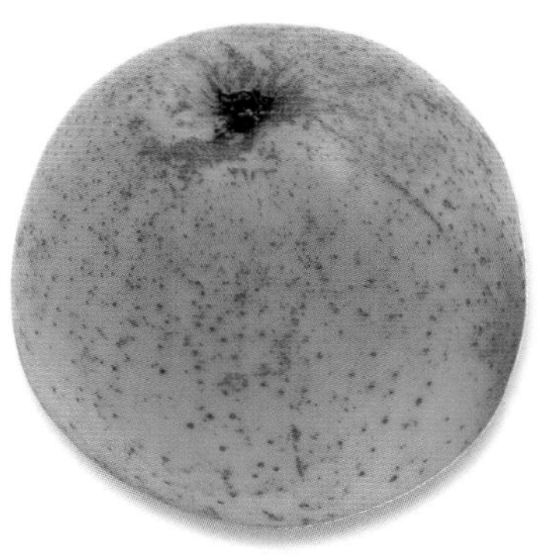

Note: The fruits can be stored for two to three months. For the best flavour, fruits should be allowed to ripen fully on the tree.

Stonetosh

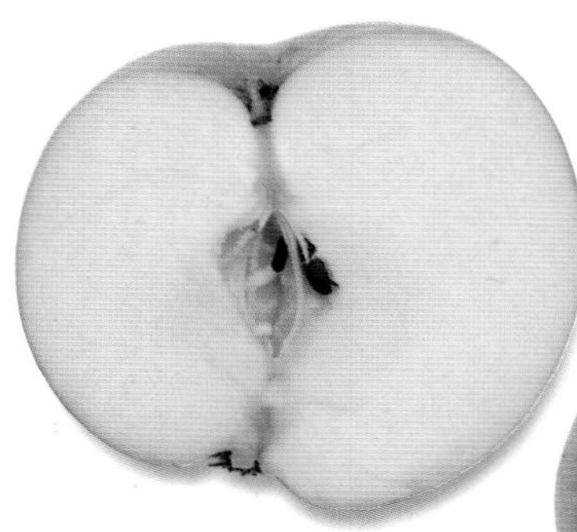

The rounded, slightly irregular fruits, around 7cm (2¾in) across, are yellowish green with a dark red flush and some streaking. The white flesh is rather soft with a moderately sweet and slightly acid flavour.

Type: Dessert.
Origin: Horticulture Division, Experimental Farm, Ottawa, Canada, 1909, introduced 1923.
Parentage: Stone (female) x McIntosh (male).
Flowering: Mid-season.
Other name: Stontosh.

Note: This variety shows some resistance to bitter pit. Trees are very vigorous.

Storey's Seedling

The rounded, slightly conical fruits, around 7cm (2¾in) across, are bright green with a strong red flush and some russeting. The white flesh is somewhat soft with a sweet and slightly subacid flavour.

Type: Dessert.
Origin: Northolt Park, Middlesex, UK, 1927.
Parentage: Newtown Pippin (female) x Unknown.
Flowering: Mid-season.

Note: Fruits can be stored throughout the winter. Trees are self-sterile so need a suitable pollination partner.

Sturmer Pippin

Type: Dessert.
Origin: Sturmer, Suffolk, UK, first recorded 1831.
Parentage: ?Ribston Pippin (female) x ?Nonpareil (male).
Flowering: Mid-season.
Other names: Apple Royal, Creech Pearmain, Moxhay, Pearmain de Sturmer, Pepin de Sturmer, Pepin iz Shturmera, Royal, Sturmer's Pepping, Sturmer, Sturmer Pepping, Sturmer's Pippin and Sturmers Pepping.

The rounded fruits, around 6cm (2¼in) across, are bright yellow-green with a red flush and brown russeting. The creamy white to yellow flesh is very firm, fine-textured and juicy with a little subacid and rich aromatic flavour.

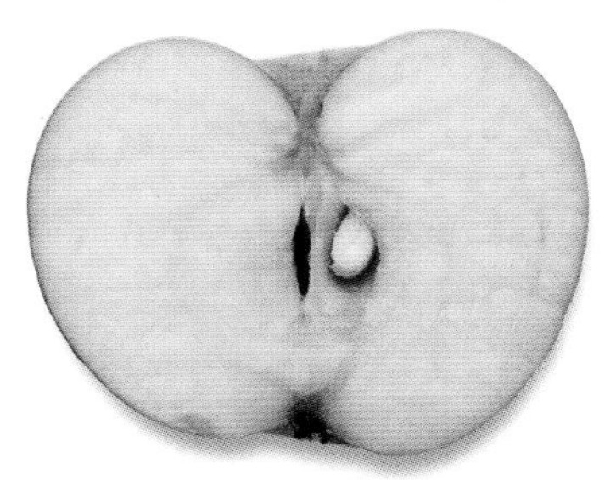

Note: Trees are very hardy and crop freely. The fruits have exceptional keeping qualities. The flavour improves in storage and can be at its best in late winter. It was taken to Australia in the 19th century, as its keeping capacity makes it useful for exporting.

Sunburn

Type: Dessert.
Origin: Hornchurch, Essex, UK, 1925.
Parentage: Cox's Orange Pippin (female) x Unknown.
Flowering: Mid-season.

The rounded fruits, around 6cm (2¼in) across, are bright yellow-green with a strong red flush. The creamy white flesh is soft with a sweet, subacid, aromatic flavour.

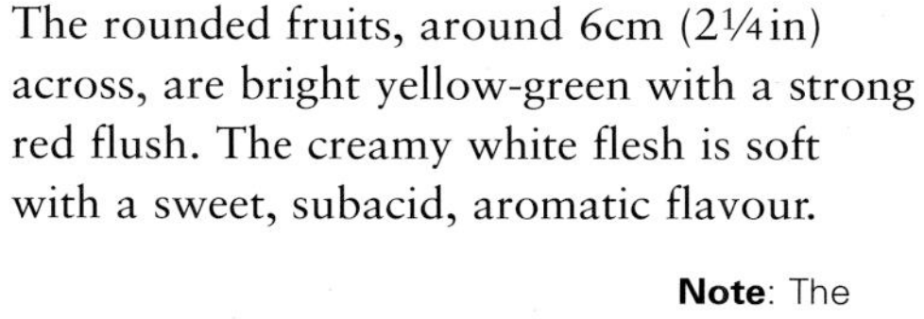

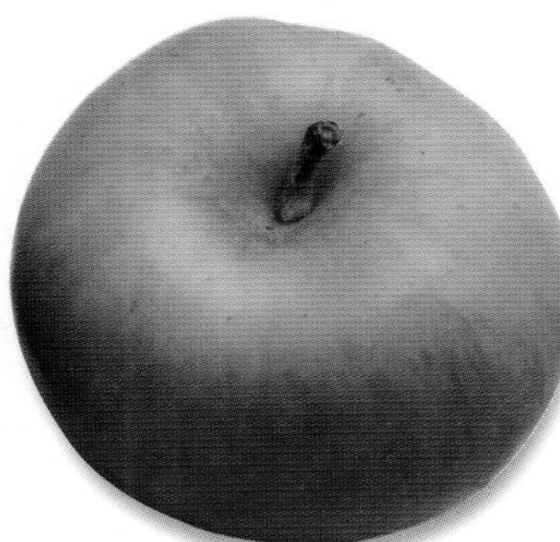

Note: The fruits can be stored for two to three months.

Right: The red flush develops on the side facing the sun.

Sunset

Type: Dessert.
Origin: Ightham, Kent, UK, *c.*1918, named 1933.
Parentage: Cox's Orange Pippin (female) x Unknown.
Flowering: Early.

The rounded, slightly flattened fruits, around 6cm (2¼in) across, are bright yellow-green with an orange-red flush and some streaking. The creamy white flesh is firm, crisp and fine-textured with a good aromatic flavour.

Note: Trees are self-fertile, show good disease resistance and crop reliably. The flavour, though variable and sometimes disappointing, is similar to Cox's Orange Pippin. The fruits can be small.

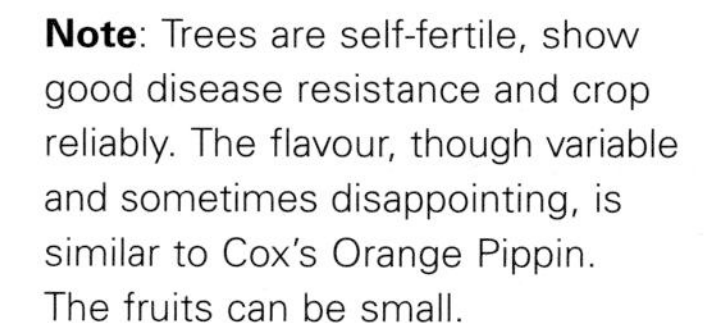

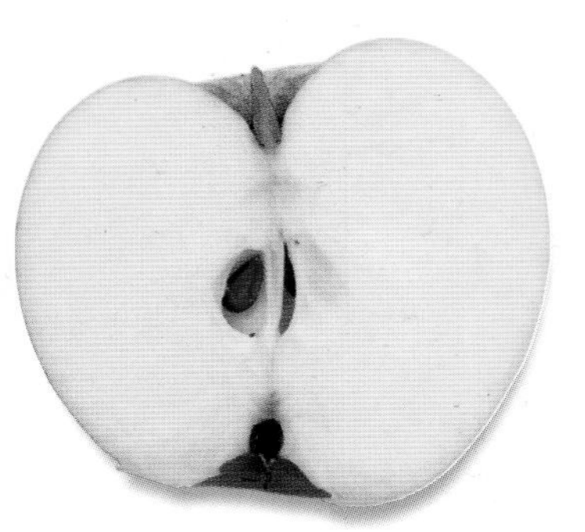

Suntan

The somewhat flattened fruits, around 6cm (2¼in) across, are bright yellow-green with a strong red flush and some streaking and patches of russeting. The creamy yellow flesh is slightly coarse-textured and moderately juicy with an aromatic, pineapple-like but acid flavour.

Type: Dessert.
Origin: East Malling Research Station, Maidstone, Kent, UK, 1956.
Parentage: Cox's Orange Pippin (female) x Court Pendu Plat (male).
Flowering: Late.

Note: This variety is prone to bitter pit. It is a triploid. Fruits are best around eight weeks after picking and can be stored for a further three months.

Swaar

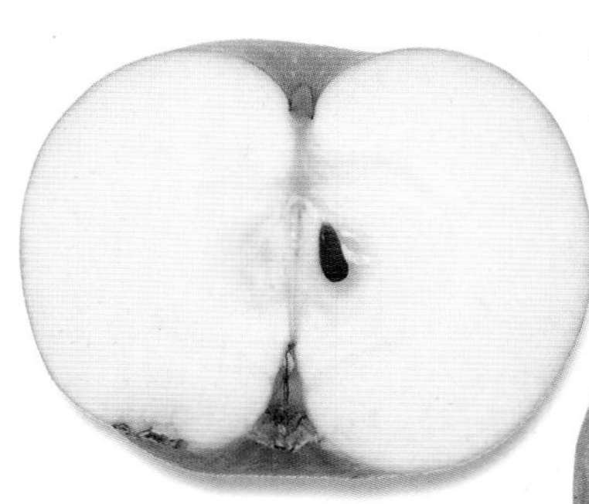

Note: The best flavour develops in storage.

The rounded, uneven fruits, around 6cm (2¼in) across, are dull yellow with russeting and spotting. The creamy white flesh is firm and fine with a sweet, aromatic, nutty flavour.

Type: Dessert.
Origin: Nr Esopus, Hudson River, NY, USA (allegedly), recorded 1804.
Parentage: Unknown.
Flowering: Mid-season.
Other names: Der Schwere Apfel, Hardwich, Hardwick, Schwere Apfel, Schwerer Apfel, Suaar, Swaar Appel, Swaar Apple and Zwaar.

Sweet-Tart

The rounded, slightly irregular fruits, around 6cm (2¼in) across, are yellow-green with a deep pinkish red flush. The creamy white flesh is soft and juicy with a very sweet flavour.

Note: Early flowering makes this variety unsuitable for growing in areas where late frosts are likely.

Type: Dessert.
Origin: Gordon Apple Trees, Whittier, CA, USA, date unknown.
Parentage: Unknown.
Flowering: Very early.

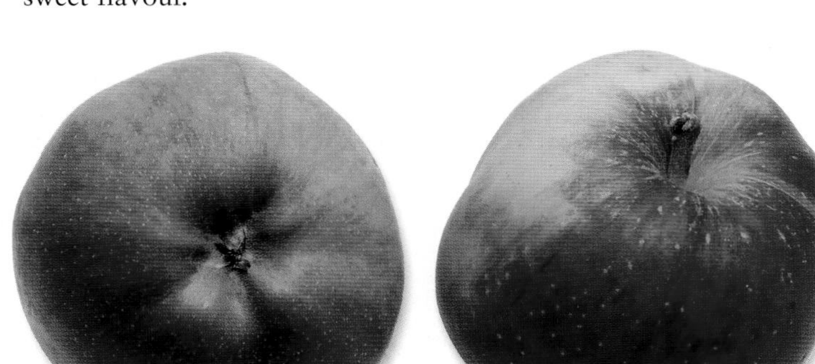

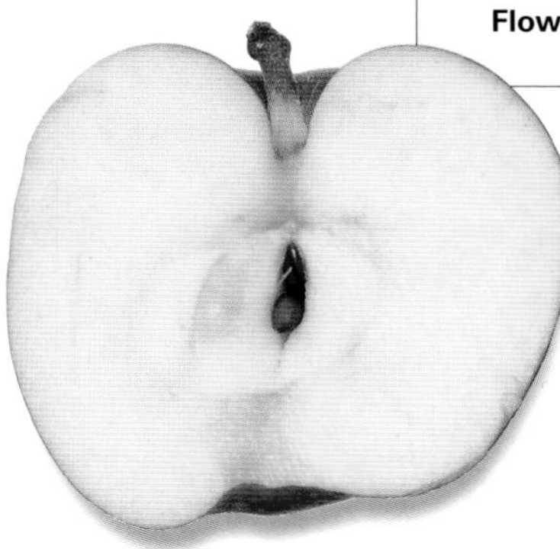

Szabadkai Szercsika

Type: Dessert.
Origin: Hungary, 1948.
Parentage: Unknown.
Flowering: Early.

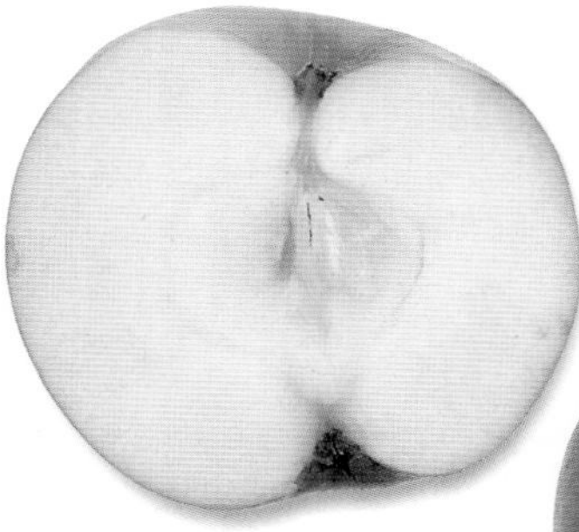

The rounded, somewhat irregular fruits, up to around 7cm (2¾in) across, are bright green with a pinkish red flush. The creamy white flesh is firm with a slightly sweet, subacid flavour.

Note: Early flowering makes this variety unsuitable for growing in frost-prone areas.

Left: Exposure to the sun brings out the characteristic pinkish flush of this variety.

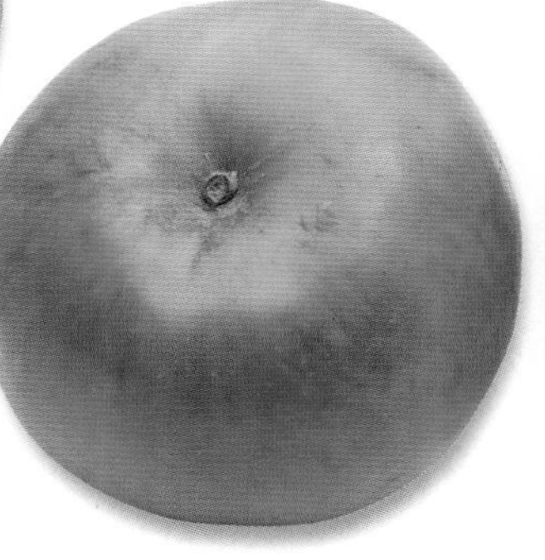

Tellina

Type: Dessert.
Origin: Italy, 1958.
Parentage: Unknown.
Flowering: Mid-season.

The conical, somewhat irregular fruits, around 6cm (2¼in) across, are bright yellow-green with a red flush and some striping and streaking. The greenish white flesh is firm and fine with a subacid flavour.

Note: Trees are moderately vigorous.

Far left: The fruits of this variety can develop an uneven shape.

Telstar

Type: Dessert.
Origin: Greytown, Wairarapa, New Zealand, 1943, named 1965.
Parentage: Golden Delicious (female) x Kidd's Orange Red (male).
Flowering: Mid-season.

The rounded fruits, around 6cm (2¼in) across or more, are dull yellow-green with a red flush and some streaking. The creamy white flesh is firm and coarse with a sweet, slightly acid and rich, complex flavour.

Note: This variety has the same parentage as Gala. Trees do not grow vigorously but crop freely.

Tenroy

The rounded fruits, around 6cm (2¼in) across or more, are bright yellow-green with a strong red flush and some flecking. The creamy white flesh is sweet, crisp and juicy with an aromatic flavour.

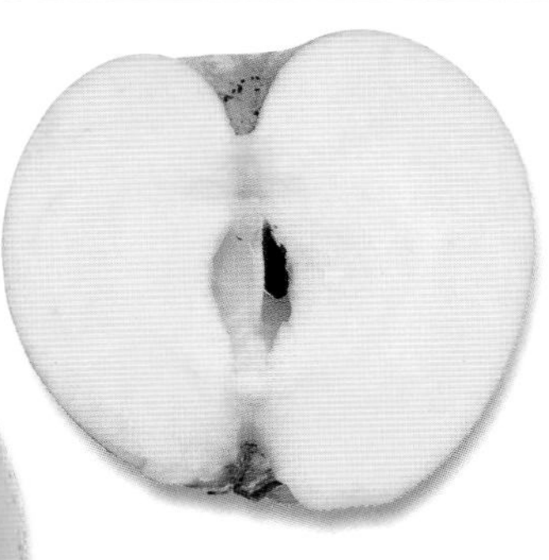

Note: This variety is a more highly coloured clone of Gala. Trees do not do well at high altitude or in dry soil. Leaves are susceptible to spotting.

Type: Dessert.
Origin: New Zealand, discovered 1971, introduced 1974.
Parentage: Kidd's Orange Red (female) x Golden Delicious (male).
Flowering: Early.

Tentation

The rounded fruits, similar to Golden Delicious, are golden yellow with an orange flush. The creamy white flesh is firm, crisp and juicy, with a pleasant, refreshing, sweet, nutty flavour.

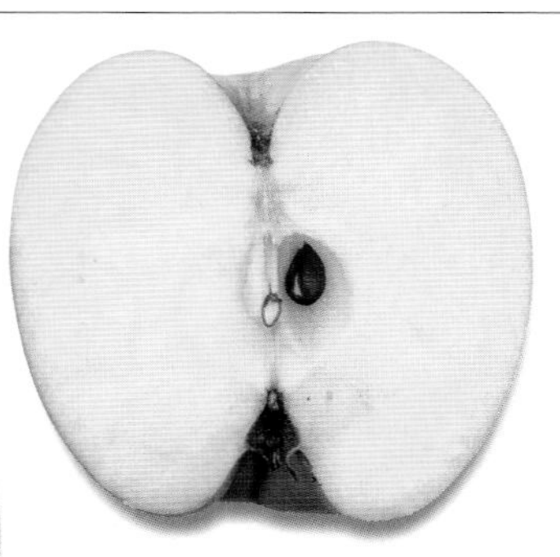

Note: This variety is potentially important commercially owing to the appeal of its colouring, which develops best in the southern hemisphere. Trees are of average vigour and set fruit at a young age.

Type: Dessert.
Origin: France, 1990 or 1979
Parentage: Golden Delicious (female) and Grifier (male).
Flowering: Mid-season.
Other name: Delblush.

Texola

The rounded, sometimes almost square, fruits, around 7cm (2¾in) across, are bright yellow-green with a red flush. The white flesh, tinged green, is soft with a subacid flavour.

Type: Dessert.
Origin: Utah, USA, 1930.
Parentage: Ben Davies (female) x Unknown.
Flowering: Early.

Note: Trees can be weak growing so may not be suitable for the most dwarfing rootstocks.

Far right: The flushing develops on the side of the fruit nearest the sun.

Tillington Court

Type: Culinary and dessert.
Origin: Burghill, Herefordshire, UK, recorded 1988 but much older.
Parentage: Unknown.
Flowering: Very early.

The rounded, slightly irregular fruits, around 7cm (2¾in) across, are bright yellow-green with a red flush which is sometimes patchy. The yellow flesh is juicy with an acid taste.

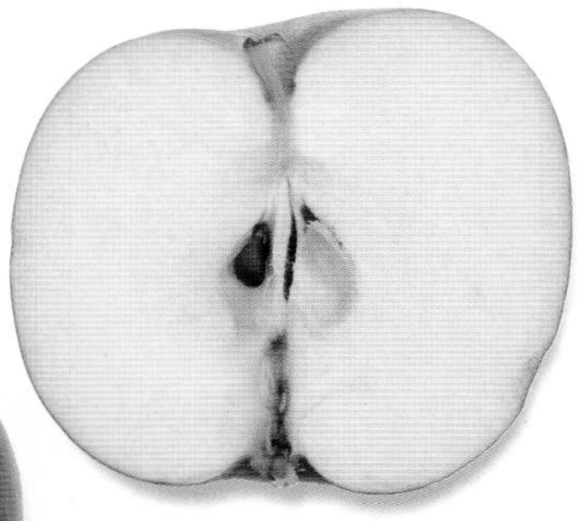

Note: Early flowering makes this variety unsuitable for growing in areas where late frosts are likely.

Tom Putt

Type: Culinary.
Origin: Trent, Somerset, UK, late 1700s.
Parentage: Unknown.
Flowering: Mid-season.
Other names: Coalbrook, Devonshire Nine Square, Izod's Kernel, January Tom Putt, Jeffrey's Seedling, Marrow Bone, Ploughman, Thomas Jeffreys, Tom Potter and Tom Put.

The uneven fruits, up to around 7cm (2¾in) across or more, are clear yellow-green with a strong crimson red flush. The creamy white flesh, stained red beneath the skin, is crisp and juicy with an acid flavour.

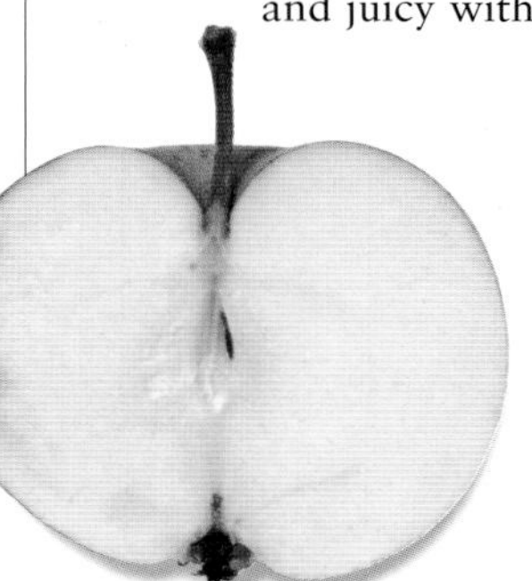

Note: The fruits, best used straight from the tree, cook well. They also make excellent cider.

Right: Fruits of this variety are often ribbed and knobbly.

Twenty Ounce

Type: Culinary and dessert.
Origin: ?New York or Connecticut, USA, 1844.
Parentage: Unknown.
Flowering: Mid-season.
Other names: Aurora, Cayuga Red Streak, Coleman, de Dix-huit Onces, de Vin du Connecticut, Dix-huit Onces, Eighteen Ounce, Gov. Seward's, Lima, Morgan's Favourite, Pomme de Vin du Connecticut, Pomme de Vingt-onces, Pomme rayée de Cayuga, Reinette de Vingt-Onces, Twenty Ounce Apple, Twenty Ounce Pippin, Wine, Wine of Connecticut and Zwanzig Unzen.

The somewhat irregular fruits, around 7cm (2¾in) across or more, are bright yellow-green with a red flush that can appear in flecks and stripes. The creamy white flesh is coarse, juicy and moderately tender with a subacid flavour.

Note: This variety is a triploid. Trees are moderately vigorous and crop well.

Tydeman's Early Worcester

Above: The bright red flush makes the Tydeman's Early Worcester a very appealing variety.

The rounded, often irregular fruits have a strong bright red flush. The white flesh is crisp, fine-textured and juicy with a good vinous flavour.

Note: Leaf spots can be a problem.

Type: Dessert.
Origin: East Malling Research Station, Maidstone, Kent, UK, raised 1929 and introduced 1945.
Parentage: McIntosh (female) x Worcester Pearmain (male).
Flowering: Mid-season.
Other names: Early Worcester, Tydeman Early, Tydeman's Early, Tydeman's Red and Tydemans Early Worcester.

Tydeman's Late Orange

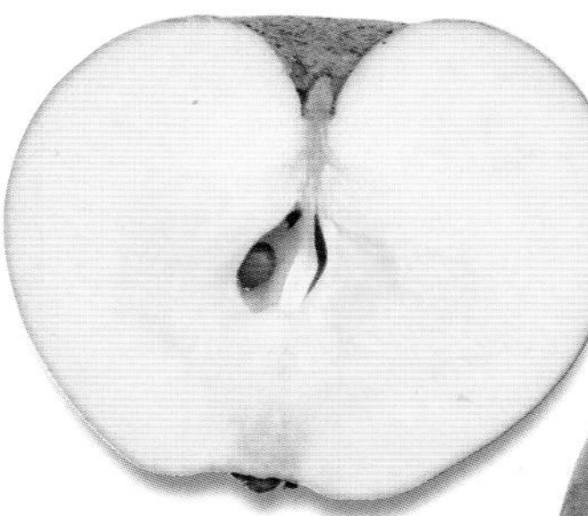

The rounded to conical fruits, around 6cm (2¼in) across, are golden green with a strong red flush and some russeting. The cream flesh is very firm, crisp and fairly juicy with a rich, aromatic flavour.

Note: The fruits store well, until early spring. The flavour is similar to a Cox but is sharper.

Type: Dessert.
Origin: East Malling Research Station, Maidstone, Kent, UK, 1930, introduced 1949.
Parentage: Laxton's Superb (female) x Cox's Orange Pippin (male).
Flowering: Mid-season.
Other name: Tydeman's Late Cox.

Underleaf

The rounded, slightly flattened fruits, around 6cm (2¼in) across, are bright yellow-green with some russeting. The creamy white flesh is sweet, with a flavour similar to Blenheim.

Note: Trees bear heavily. The fruits are suitable for cider making.

Type: Dessert.
Origin: Long Ashton Research Station, Bristol, UK, 1967.
Parentage: Unknown.
Flowering: Late.
Other name: Gloucestershire Underleaf.

Upton Pyne

Type: Culinary and dessert.
Origin: Topsham, Devon, UK, introduced 1910.
Parentage: Unknown.
Flowering: Mid-season.

The rounded fruits, around 6cm (2¼in) across or more, are bright yellow-green with a red flush and some flecking. The creamy white flesh is firm, rather coarse-textured and juicy with a somewhat acid and fair flavour.

Note: The fruit cooks to a purée. The flavour is like that of a pineapple. Trees show good disease resistance.

Left: Upton Pyne is a versatile, dual-purpose apple.

Venus Pippin

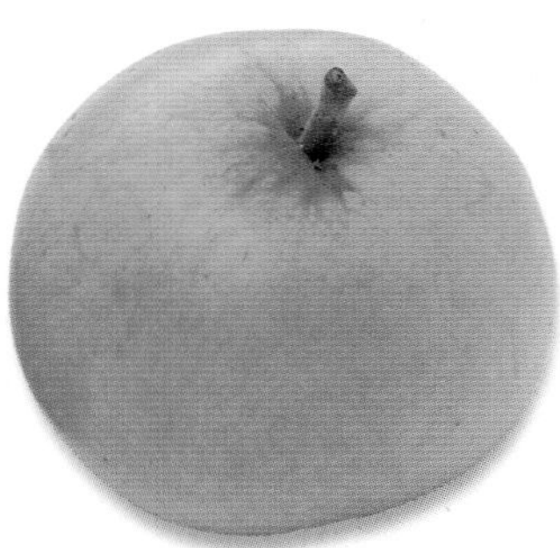

The rounded to slightly conical or rectangular fruits, around 6cm (2¼in) across, are clear bright green with a faint pinkish brown flush and some spotting. The creamy white flesh is tender, soft and coarse with a slightly sweet, slightly acid flavour.

Type: Culinary.
Origin: Thought to be Devon, UK, *c.*1800.
Parentage: Unknown.
Flowering: Mid-season.
Other names: Plumderity and Venus' Pippin.

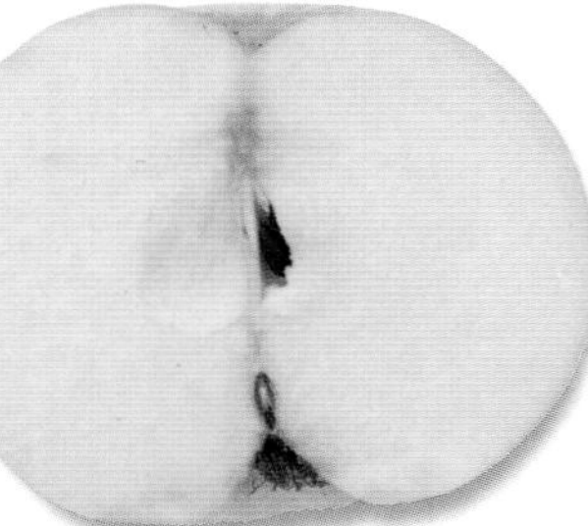

Note: Trees are very vigorous so are best grown on very dwarfing rootstocks.

Right: The clear, almost yellowish, colouring of the ripe fruits is unusual.

Verallot

Type: Culinary.
Origin: France, 1948.
Parentage: Unknown.
Flowering: Very late.

The flattened fruits, around 6cm (2¼in) across, are bright yellow-green with a bright red flush and some russeting. The greenish white flesh is firm with an acid flavour.

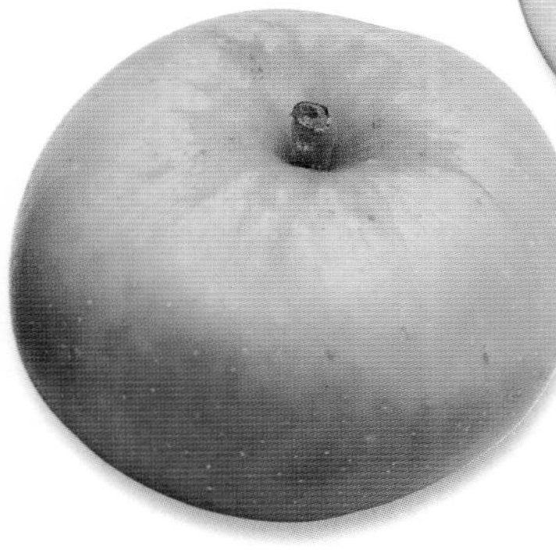

Note: Late flowering makes this a very useful variety for growing in areas where late frosts are likely to occur.

Vérité

The somewhat flattened fruits, around 6cm (2¼in) across, are bright yellow-green with a strong red flush. The greenish white flesh is firm with a slightly sweet, subacid flavour.

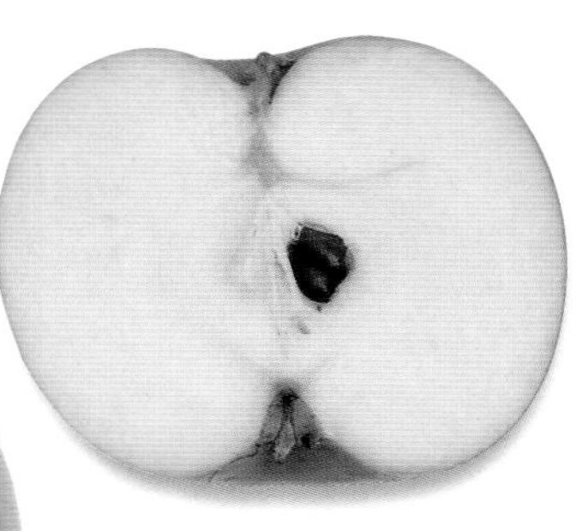

Type: Dessert.
Origin: France, recorded 1876.
Parentage: Unknown.
Flowering: Very late.

Note: Fruits, which store well, can also be used in cooking and are good for juicing and cider making. Trees are very hardy.

Vernajoux

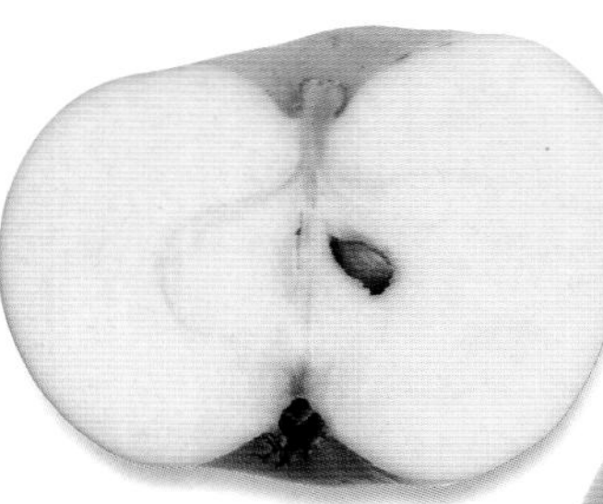

The rounded, occasionally somewhat flattened fruits, around 6cm (2¼in) across, are bright yellow-green with a light red flush and some russeting. The white flesh, tinged green, is firm and tough with a slightly sweet flavour.

Type: Dessert.
Origin: France, described 1947.
Parentage: Unknown.
Flowering: Late.

Note: Fruits can be stored for around four months, sometimes longer. They show some resistance to bruising.

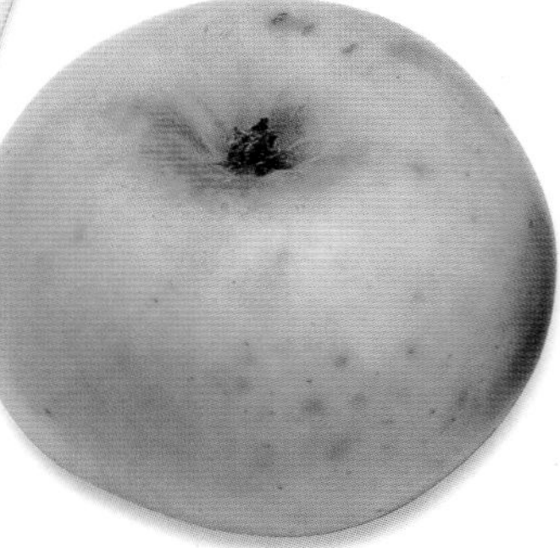

Victory

The rounded to slightly conical fruits are around 6cm (2¼in) across. They are green with a pinkish red flush and some darker flecking. The creamy white flesh is firm with an acid flavour.

Note: This variety should not be confused with the American Victory, normally styled as Victory (USA).

Type: Culinary.
Origin: Unknown.
Parentage: Bismarck (female) x Blenheim Orange (male).
Flowering: Mid-season.
Other name: Carpenter.

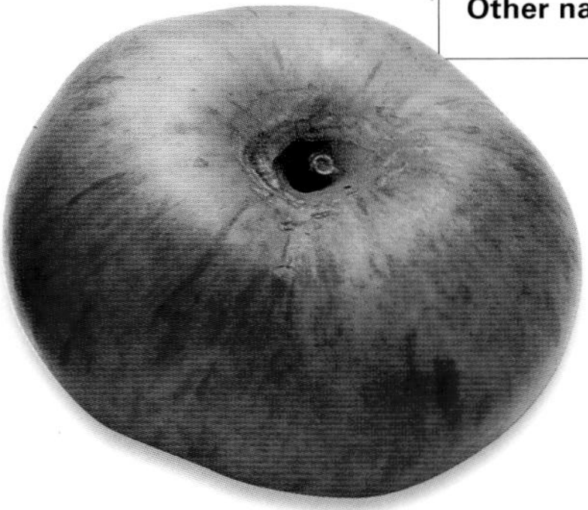

Vincent

Type: Dessert.
Origin: France, described 1947.
Parentage: Unknown.
Flowering: Late.
Other name: Saint-Vincent.

The slightly flattened fruits, around 6cm (2¼in) across, are yellowish green with a red flush and russeting. The creamy white flesh is firm and crisp with a subacid and slightly sweet flavour.

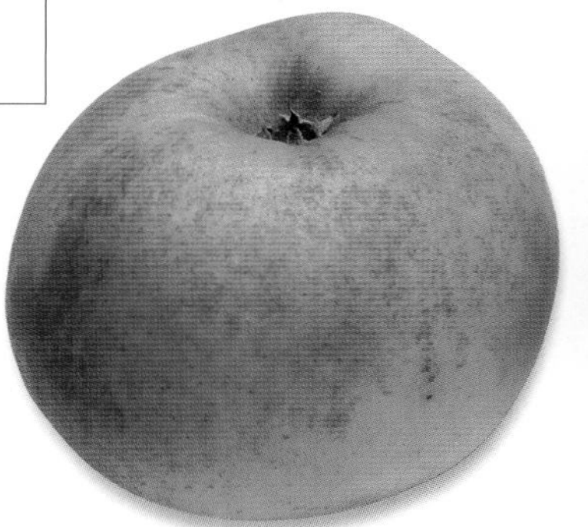

Note: Trees are moderately vigorous. Late flowering makes this a suitable variety for areas where late frosts are likely.

Violetta

Type: Dessert.
Origin: Italy, 1958.
Parentage: Unknown.
Flowering: Mid-season.

The rounded fruits, around 5cm (2in) across, are bright yellow-green with a bright red flush and some flecking. The creamy white flesh is firm and fine with a subacid and slightly sweet flavour.

Note: This variety is unusual in commerce outside its country of origin.

Far left: The flush on the skins deepens on the area nearest the sun.

Violette

Type: Dessert.
Origin: France, known early 1600s.
Parentage: Unknown.
Flowering: Early.
Other names: Black Apple, Calville Rayée d'Automne, De Quatre-Goûts, de Violette, Des Quatre-Goûts, Framboise, Framboox Apple, Grosse Pomme Noire d'Amérique, Pomme Violette, Quatre Goûts, Reinette des Quatre-Goûts, Reinette Violette, Violet, Violette de Mars, Violette de Quatre Goûts, Violetter Apfel and Winter Veilchen.

The conical to oblong, somewhat irregular fruits, around 6cm (2¼in) across, are bright yellow-green with a pinkish to very dark, almost bluish to blackish, red flush. The yellowish white flesh, tinged red under the skin, is firm and fine with a subacid and slightly aromatic flavour.

Note: This variety shows some resistance to black spot. Trees are very vigorous. The fruits are also suitable for cooking.

Wagener

The rounded, very slightly flattened, fruits, around 6cm (2¼in) across, are bright green with a strong red flush. The creamy white flesh is firm and moderately juicy with a pleasant flavour.

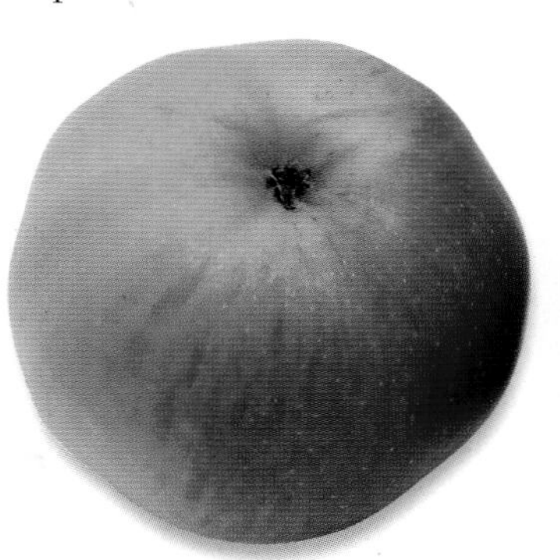

Note: Trees are hardy, scab-resistant and bear heavily. Thinning may be necessary. The fruits can also be used for apple sauce and in cider making. They do not shrivel in storage.

Type: Dessert.
Origin: Penn Yann, NY, USA, 1791.
Parentage: Unknown.
Flowering: Early.
Other names: Pomme Wagener, Vagner, Vagnera Premirovannoe, Vagnera Prizovoe, Wagener Premiat, Wagener Price Apple, Wagener's, Wagenerapfel, Wageners Preisapfel, Waggoner, Wagner, Wagner Dijas, Wagner Preiss Apfel, Wagner Premiat, Wagnera Prizovoe and Wegenerovo.

Wang Young

The rounded fruits, around 6cm (2¼in) across, are bright green with a pinkish to dark red flush. The creamy white flesh is soft, juicy and sweet.

Note: Trees are moderately vigorous. This is one of the few South Korean apples in commerce outside its country of origin.

Type: Dessert.
Origin: South Korea, 1967.
Parentage: Unknown.
Flowering: Mid-season.

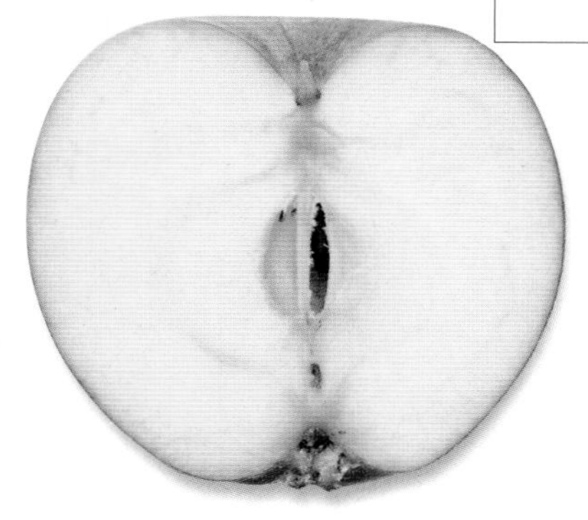

Warden

The rounded fruits, around 6cm (2¼in) across, are bright yellow with a strong dark red flush and flecking. The white flesh is soft and juicy with a fairly sweet flavour.

Note: Trees are moderately vigorous. Warden is an old English term for a pear.

Type: Dessert.
Origin: Scotland, UK, 1967.
Parentage: Unknown.
Flowering: Mid-season.

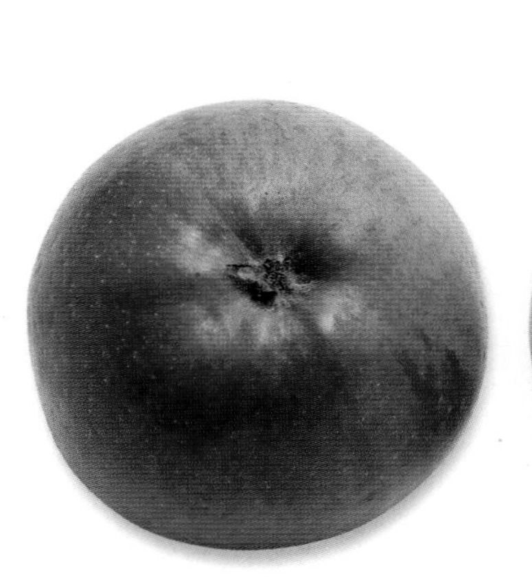

Wealthy

Type: Dessert.
Origin: Excelsior, MN, USA, first recorded 1860.
Parentage: Cherry Crab (female) x Unknown.
Flowering: Mid-season.
Other names: Lelsy, Plodovodnoe and Uelsi.

The flattened fruits, around 6cm (2¼in) across, are bright yellow-green with a pinkish red flush and some streaking. The creamy white flesh is rather soft, coarse-textured and juicy with a sweet and faintly vinous flavour.

Note: Trees bear heavily, even when young, but are prone to biennial bearing. A long flowering period makes this a good pollinator for many other varieties.

Right: This was the first apple variety of commercial quality to be grown in Minnesota and eventually became one of the five most-produced apples in the USA.

Wellington Bloomless

Type: Dessert.
Origin: USA, 1966.
Parentage: Unknown.
Flowering: Mid-season.

The irregular, quince-like fruits, around 7cm (2¾in) across, are yellow-green with a strong red flush and some spotting. The creamy white flesh has a sweet and pleasant flavour.

Above: No two fruits of this variety are likely to show the same shape.

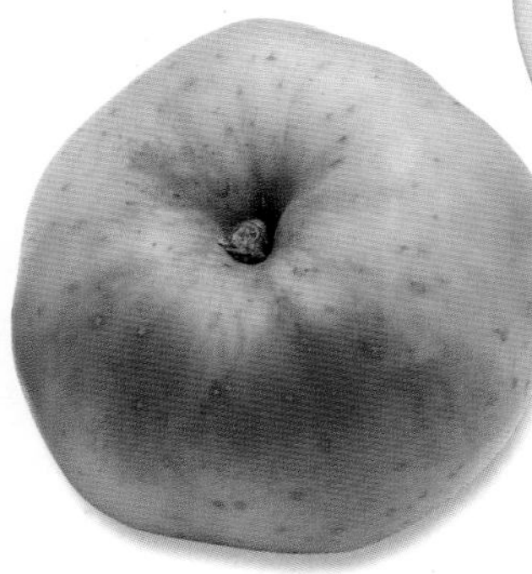

Note: A unique feature of this variety is that its flowers have no petals – hence the name.

Wellspur Delicious

Type: Dessert.
Origin: Azwell Orchard of the Wells and Wade Fruit Company, Wenatchee, WA, USA, discovered 1952, introduced 1958.
Parentage: Unknown.
Flowering: Mid-season.
Other name: Wellspur.

The rounded but irregular fruits, around 6cm (2¼in) across, are bright green with a strong pinkish to dark red, sometimes patchy, flush. The creamy white flesh is very firm, very sweet and juicy with a highly aromatic flavour.

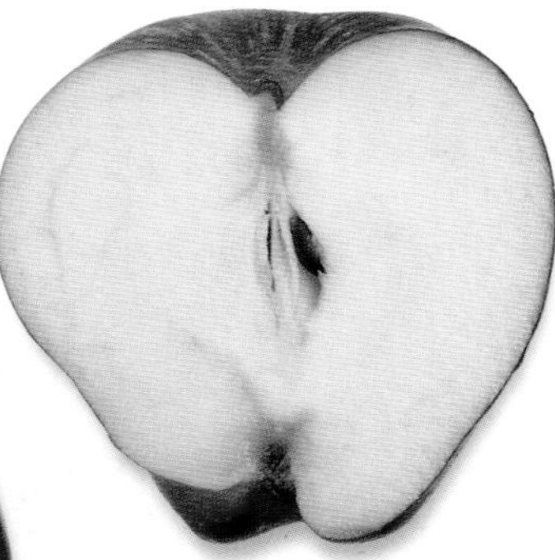

Note: This variety is a sport of Starking with a more solid red flush.

Wheeler's Russet

The flattened, irregular fruits, around 6cm (2¼in) across, are mid-green with yellowish grey to reddish brown russeting and some freckling. The greenish white flesh is a little soft with a subacid, slightly sweet and slightly aromatic flavour.

Type: Dessert.
Origin: England, UK, known in 1717.
Parentage: Unknown.
Flowering: Mid-season.
Other name: Reinette Grise de Wheeler.

Note: Trees are very hardy and bear well. The fruits can be stored for four to five months. They retain a good flavour even as they begin to shrivel.

White Winter Pearmain

The irregular fruits, around 6cm (2¼in) across, are yellowish green with a red flush and some dotting. The creamy white flesh is firm, crisp, tender and fine with a subacid and aromatic flavour.

Type: Dessert.
Origin: Thought to have originated in the eastern states, USA, 1867.
Parentage: Unknown.
Flowering: Mid-season.

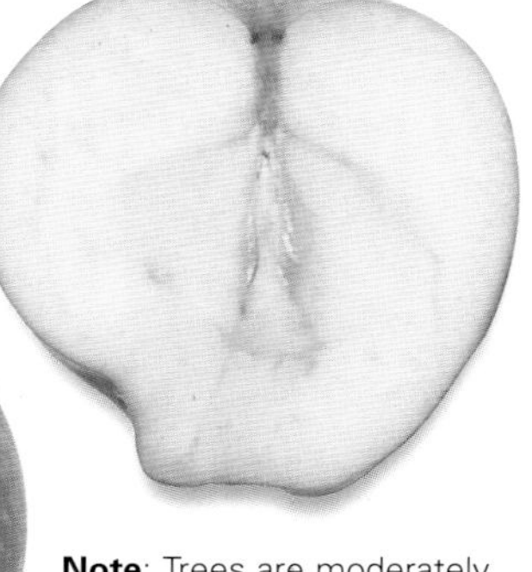

Note: Trees are moderately vigorous. This variety makes an excellent pollinator for other apple trees.

Widdup

The rounded fruits, up to around 7cm (2¾in) across, are bright green with a red flush and some flecking. The greenish white flesh is firm, fine and crisp with a subacid and slightly sweet flavour.

Note: This variety is increasingly rare in cultivation.

Type: Dessert.
Origin: New Zealand, 1961.
Parentage: Unknown.
Flowering: Early.

William Crump

Type: Dessert.
Origin: Rowe's Nurseries, Worcester, UK, 1908.
Parentage: Cox's Orange Pippin (female) x Worcester Pearmain (male).
Flowering: Mid-season.

The rounded, slightly irregular fruits, around 6cm (2¼in) across or more, are yellow-green with a strong red flush. The creamy white flesh is firm, fine-textured and juicy with a sweet and rich, aromatic flavour.

Note: Trees are very vigorous. The fruits can be stored for two to three months.

Right: *The regular flecking on the skins makes this a very appealing apple.*

Winesap

Type: Culinary.
Origin: USA, first described 1817.
Parentage: Unknown.
Flowering: Mid-season.
Other names: American Winesop, Banana, Henrick's Sweet, Holland's Red Winter, Pot Pie Apple, Red Sweet Wine Sop, Royal Red, Royal Red of Kentucky, Texan Red, Virginia Winesaps, Winesopa and Winter Winesap.

The rounded fruits, up to around 7cm (2¾in) across, are dull yellow-green with a pinkish red flush. The yellowish white flesh is firm, tender and coarse with a sweet, subacid flavour.

Note: The fruits store well. They are suitable for cider making. The skins can be tough.

Winston

Type: Dessert.
Origin: Welford Park, Berkshire, UK, introduced 1935 as Winter King, renamed 1944.
Parentage: Cox's Orange Pippin (female) x Worcester Pearmain (male).
Flowering: Mid-season.
Other names: Cox d'Hiver, Winter King and Wintercheer.

The rounded to slightly conical fruits, around 6cm (2¼in) across, are bright yellow-green with a strong red flush and some russeting. The creamy white flesh is firm, fine-textured and juicy with a sweet and good aromatic flavour.

Note: Trees are resistant to most diseases and crop reliably.

Right: *Fruits of this variety tend to taper towards the blossom end.*

Winter Banana

The irregular fruits, up to around 7cm (2¾in) across or more, are pale yellow-green with a red flush. The creamy white flesh is rather soft, rather coarse-textured and moderately juicy with a sweet and pleasant aromatic flavour.

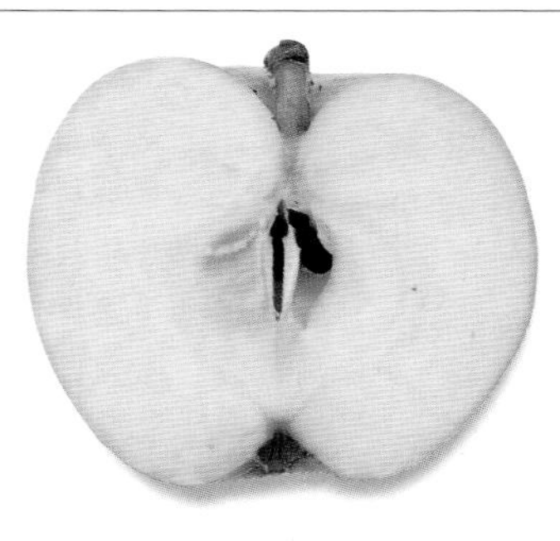

Note: This variety is suitable for training, especially as an espalier. Trees, which are vigorous, are resistant to nearly all diseases. The flavour is similar to that of a banana. The fruits can also be cooked.

Type: Dessert.
Origin: Cross County, IN, USA, introduced 1890.
Parentage: Unknown.
Flowering: Mid-season.
Other names: Banan zimnii, Banana, Banana de Iarna, Banane, Banane d'Hiver, Bananove zimni, Bananovoe, Flory, Teli banan and Zimna bananova.

Winter Peach

The flattened, somewhat irregular fruits, up to around 7cm (2¾in) across or more, are pale green with a red flush. The yellowish flesh is crisp, tender and juicy with an acid and slightly spicy flavour.

Note: The fruits can be stored for up to six months or even longer. They are also suitable for cooking.

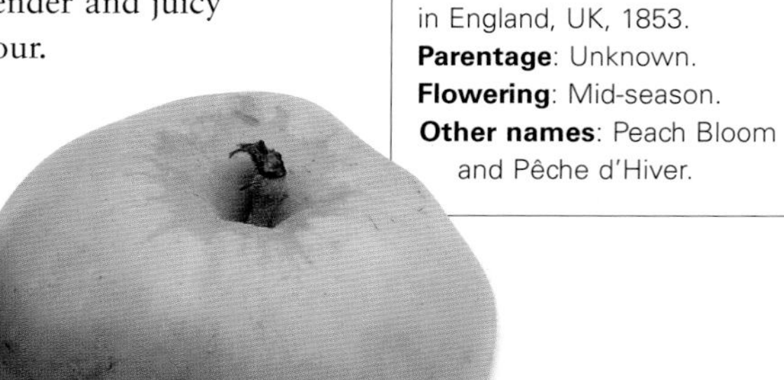

Type: Dessert.
Origin: ?USA, known in England, UK, 1853.
Parentage: Unknown.
Flowering: Mid-season.
Other names: Peach Bloom and Pêche d'Hiver.

Winter Pearmain

The rounded to conical fruits, up to around 7cm (2¾in) across, are bright green with a strong red flush and some russeting and dotting. The yellowish flesh is firm and crisp with a pleasant, sweet, subacid flavour.

Note: Trees are very hardy and bear freely. The fruits can be stored for three to four months.

Type: Culinary and dessert.
Origin: Thought to be an old English variety, date unknown.
Parentage: Unknown.
Flowering: Early.
Other names: Duck's Bill, English Winter Pearmain, Grange's Pearmain, Grauwe of Blanke Pepping van der Laan, Great Pearmain, Hertfordshire Pearmain, Old Winter Pearmain, Pepin Pearmain d'Angleterre, Pepin Pearmain d'Hiver, Reinette très tardive, Somerset Apple Royal, Striped Winter, Sussex Scarlet Pearmain, Sussex Winter Pearmain and Winter Queening.

Woolbrook Pippin

Type: Dessert.
Origin: Sidmouth, Devon, UK, 1903.
Parentage: Cox's Orange Pippin (female) x Unknown.
Flowering: Mid-season.

The irregular fruits, up to around 7cm (2¾in) across, are bright yellow-green with a pinkish red flush and some striping. The creamy white flesh is firm, crisp and tender with a sweet, slightly acid and aromatic flavour.

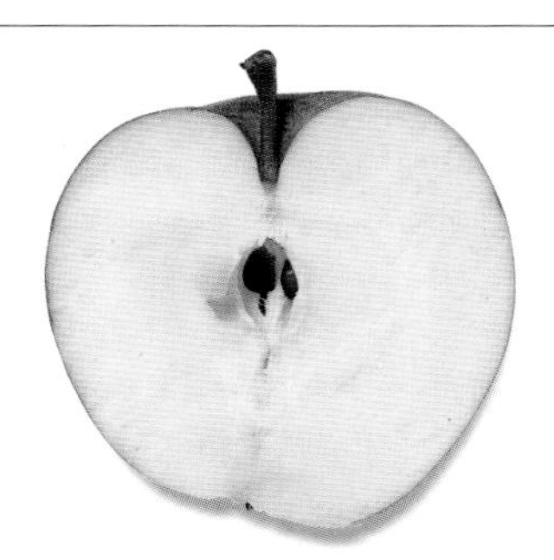

Note: This variety shows some resistance to scab and canker. Trees are vigorous and upright.

Right: This attractive variety is widely used in heritage plantings of older varieties.

Worcester Pearmain

Type: Dessert.
Origin: Swan Pool, near Worcester, UK, introduced 1874.
Parentage: ?Devonshire Quarrenden (female) x Unknown.
Flowering: Early.
Other names: Pearmain de Worcester, Worcester, Worcester Parmaene, Worcester Parman, Worcester Parmane, Worcester-Parman and Worchester Parmane.

Note: This variety is often used in breeding programmes to develop new varieties. It makes a good garden tree, bearing freely. Fruits ripen early in the season but for the best flavour should be left as long on the tree as possible.

The rounded to conical, slightly irregular fruits, around 6cm (2¼in) across, are bright yellow-green with an intense red-crimson flush and some spotting, flecking and russeting. The white flesh is firm and a little juicy with a sweet and pleasant, sometimes strawberry-like flavour.

Wyken Pippin

Type: Dessert.
Origin: ?Wyken, nr Coventry, UK; also said to have been introduced from Holland, early 1700s.
Parentage: Unknown.
Flowering: Mid-season.
Other names: Airley, Alford Prize, Arley, Gerkin Pippin, German Nonpareil, Girkin Pippin, Pepin de Warwickshire, Pepin du Warwick, Pepping aus Warwickshire, Pepping von Wyken, Pheasant's Eye, Pippin du Warwick, Warwick Pippin, Warwickshire Pippin and White Moloscha.

The rounded to flattened fruits, around 6cm (2¼in) across, are bright yellow-green with a dull orange flush and russeting that appears in spots. The creamy white flesh, tinged with green, is moderately firm, fine-textured and very juicy with a sweet and good aromatic flavour.

Note: Trees are generally healthy and crop freely. The fruits can be stored for two to three months.

Yellow Bellflower

The conical, somewhat irregular fruits, around 6cm (2¼in) across or more, are bright yellow-green with an orange-red flush and some spotting. The creamy white flesh is moderately firm and crisp with a sweet, slightly subacid flavour.

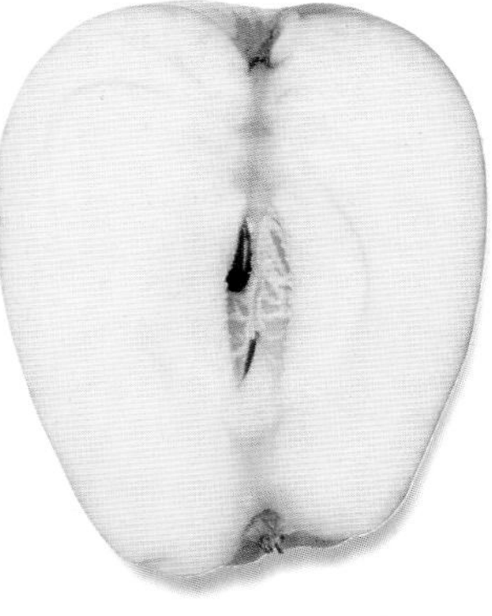

Note: Trees, which are vigorous and quick-growing, are suitable for training. They are resistant to leaf spots.

Type: Culinary and dessert.
Origin: Burlington County, NJ, USA, by 1817.
Parentage: Unknown.
Flowering: Early.
Other names: Belfiore giallo, Belfler zheltyi, Bell Flower, Belle Fleur jaune, Belle Flavoise, Belle Fleur, Belle Flower, Belle-Flavoise, Belle-Fleur Yellow, Belle-Flower, Bellefleur Yellow, Bellflower, Bishop's Pippin, Bishop's Pippin of Nova Scotia, Calville Metzger, Connecticut Seek-no-Further, Frumos galben, Gelber Bellefleur and Gelber Englischer Schönblühender.

Yellow Ingestrie

The rounded to square fruits, around 6cm (2¼in) across, are bright yellow-green sometimes tinged with a deeper yellow. The greenish yellow flesh is fine and tender with a rich and subacid vinous flavour.

Note: Trees are large and spreading and crop abundantly.

Left: This old English variety has very distinctive coloration.

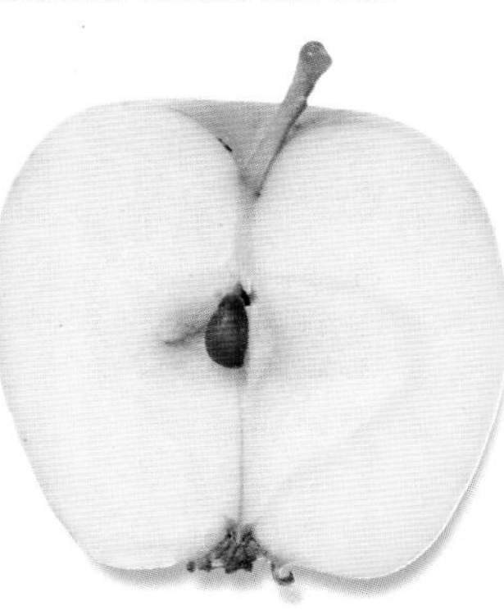

Type: Dessert.
Origin: Wormsley Grange, Herefordshire, UK, *c.*1800.
Parentage: Orange Pippin (female) x Golden Pippin (male).
Flowering: Early.
Other names: d'Ingestrie jaune, Early Pippin, Early Yellow, Gelber Pepping von Ingestrie, Ingestrie, Ingestrie Jaune, Ingestrie Yellow, Little Golden Knob, Pomme d'Ingestrie Jaune, Summer Golden Pippin, White Pippin and Yellow Ingestrie Pippin.

Yorkshire Greening

The flattened, very irregular fruits, up to around 7cm (2¾in) across or more, are bright green with a red flush and some striping. The creamy white flesh is firm, rather coarse-textured and somewhat dry with a very acid flavour.

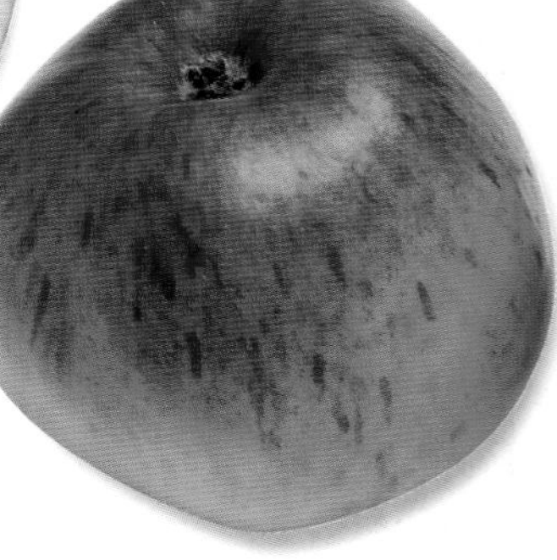

Note: Trees are very hardy, performing well in difficult locations, and crop reliably. The fruits, which can be stored for up to two or three months, cook to a purée.

Type: Culinary.
Origin: ?Yorkshire, UK, recorded 1803.
Parentage: Unknown.
Flowering: Early.
Other names: Coate's, Coates, Coates', Coates' Greening, Goose Sauce, Grünling von Yorkshire, Seek-no-Farther, Seek-no-Further, Verte du Comte d'York, Yorkshire Goose Sauce and Yorkshire Greeting.

Zabergäu Renette

Type: Dessert.
Origin: Hausen an der Zaber, Baden-Württemberg, Germany, from seed sown in 1885.
Parentage: Unknown.
Flowering: Mid-season.
Other names: Graue Renette von Zabergäu, Hausener Graue Renette, Zabergäu, Zabergäu-Renette and Zabergäurenette.

The rounded fruits, around 6cm (2¼in) across, are bright yellow-green with heavy russeting over the whole surface. The creamy white flesh is firm, moderately fine-textured and fairly juicy with a rich, nutty flavour.

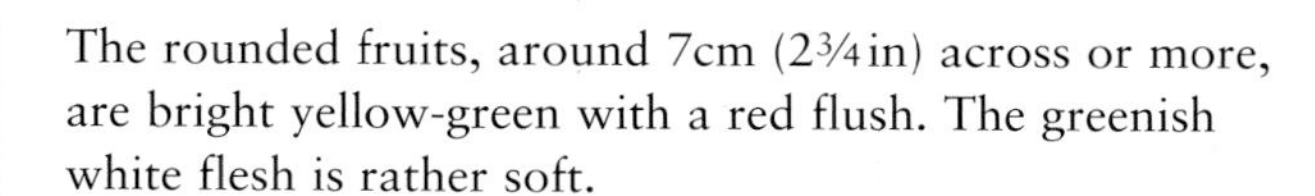

Note: This variety is a triploid. Fruits eaten straight from the tree taste of nettles. They can be stored for three months or more. Trees show good disease resistance.

Right: Fruits of this variety are notable for their uniform shape.

Zelyonka Kharkovskaya

Type: Dessert.
Origin: Russia, first half of the 1800s.
Parentage: Unknown.
Flowering: Mid-season.
Other names: Kharkovskaya zelyonka, Kharkowskaia zelenka, Zelenka, Zelenka Kharkovskaya and Zelyonka.

The rounded fruits, around 7cm (2¾in) across or more, are bright yellow-green with a red flush. The greenish white flesh is rather soft.

Note: This variety is not widely grown outside its country of origin but has been used in research programmes.

Zoete Ermgaard

Type: Dessert.
Origin: Netherlands, known since 1864.
Parentage: Unknown.
Flowering: Late.

Note: This variety is very hardy, so is suitable for growing in areas that experience hard frosts.

The slightly conical, rather uneven fruits, around 6cm (2¼in) across or more, are pale yellow-green with a strong pinkish red flush. The creamy white flesh is firm, crisp, coarse and dry with a sweet, slightly subacid flavour.

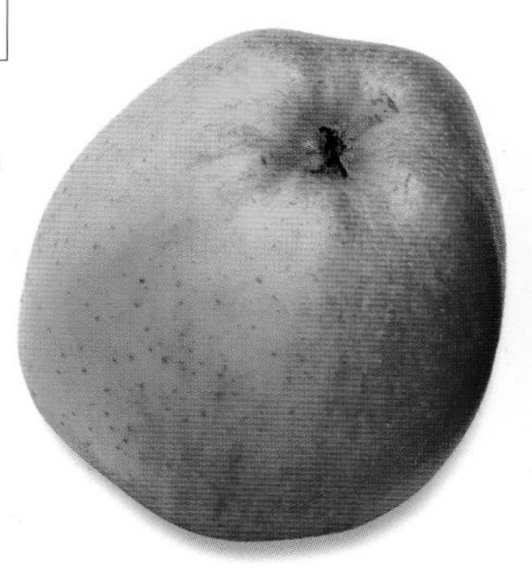

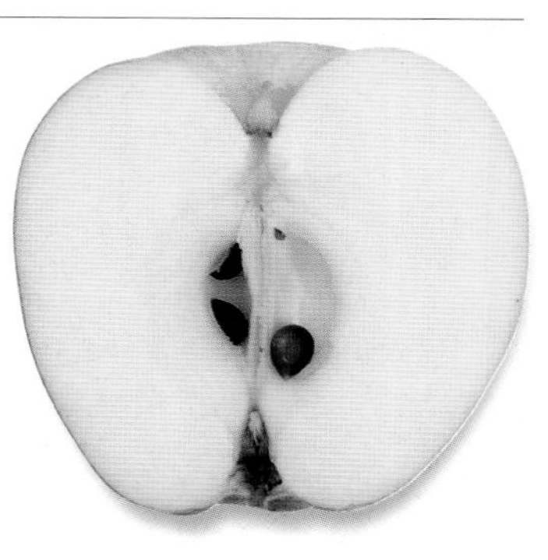

INDEX

Below: *Jonagored.*

Below: *Rossie Pippin.*

Below: *May Queen.*

Below: *Winston.*